INTRODUCTION TO PROBABILITY

Nelson G. Markley

SIMON & SCHUSTER CUSTOM PUBLISHING

Printed in the United States of America

10 9 8 7 6 5 4 3

ISBN 0-536-58737-X
BA 96386

SIMON & SCHUSTER CUSTOM PUBLISHING
160 Gould Street/Needham Heights, MA 02194
Simon & Schuster Education Group

Preface

To the First Edition:

These notes grew out of my work on Math 111 at the University of Maryland. Probability had been the major topic in this course for many years when I became involved in teaching and overseeing it. I brought no commitment to the subject or syllabus; in fact, I would not have hesitated to argue for drastic changes. Within a year I was convinced that not only should probability be the major topic but it should also be a very evident unifying theme running through the entire course.

Since statistics has been one of the major forces in the quantification of business administration and the life and social sciences, the value of a good background in probability has increased for students in these areas. The subject also has a general value. As a people we love statistical information, but are frequently ill-equipped to separate the impressive facts from the misleading ones. For the person with no formal training in statistics, an understanding of the elements of probability provides a basis for intelligently reacting to the constant stream of statistical facts in our everyday lives. These notes are aimed at taking the student through the fundamentals of probability to the edge of statistics.

An interesting feature of elementary probability is its reliance on the careful use of language. I can think of no other core area of mathematics in which a student's verbal abilities are as valuable as their mathematical aptitudes. Because these notes were written for students who tend to be more verbally oriented than those in the physical sciences, I have tried to take full advantage of this aspect of the subject. There are no formal proofs. Most of the problems are what are commonly called reading problems, and the only prerequisite is a year of high school algebra.

The exercises are an essential feature of these notes. Students can no more learn mathematics by listening to a lecture than they can learn to play the piano by going to a recital. There are over 300 exercises including review problems at the end of each chapter. I can not overstate the importance of students working and reworking as many of these exercises as possible. I have tried to include a rich assortment of problems in which the student must first recognize what methods are applicable and then have the technical skill to complete the solution. Answers to the odd numbered exercises are provided in the back of the book.

With the writing of these notes I incurred many debts of gratitude. First I want to thank Brit Kirwan for giving me the opportunity to work on Math 111 as a long term project. During this period Jerry Sather, Jim Schafer, and Scott Wolpert helped me run the course for one or more semesters. Many of the exercises are a direct product of the exam writing we did each semester. There were also many graduate students who took an interset in teaching the course. In particular, Roy Dahl and Charles Toll, both of whom had already taught a number of times, volunteered to teach it again using an experimental version of these notes. J. (Shep) Shepherd checked the text line by line and worked every exercise. His work more than anyone else's substantially reduced the number of minor errors. Special thanks go to my wife Pat. In addition to typing a superb photo ready copy, she performed invaluable editorial work on the manuscript.

College Park, Maryland N.G.M.
April 1982

To the Second Edition:

Last summer for the first time in ten years I taught elementary probability, and I enjoyed it as much as I did during the years that I wrote the first edition of this book. Once again it was refreshing to teach a beginning mathematics course with more room than most for dialogue with the students. The students did their part and asked many intelligent questions which required articulate answers from me and stimulated class discussion. Of course, there was a full range of student performance, but in general I felt the student engagement and attitude was very good.

Not surprisingly, I continue to believe elementary probability is the kind of mathematics we should be teaching to undergraduates in the humanities and the social sciences. For many of them it is fresh material with a visible linkage to the world around them. It provides those of us who teach mathematics with an excellent opportunity to give non-science students an understanding of mathematical reasoning with a minimum of symbolic manipulation. Although the prerequisites for elementary probability are minimal, there is still an important story to tell that begins with the simple idea of counting and ends with linking the theoretical and empirical notions of probability.

The impetus to publish a second edition came from Howard Community College. They share my views on the value of teaching probability to beginning college students and have recently decided to make it the foundation of their introductory course. I am particularly grateful to Bernadette Sandruck, who teaches there, for her many valuable suggestions on both updating the text and making it more usable for the students. I am also indebted to Joe Auslander, University of Maryland; Bart Braden, Northern Kentucky University; and Thomas Sonnabend, Montgomery College, for a variety of general and specific ideas which were helpful to me.

The most visible change between the two editions is the typesetting. The new edition has been prepared using Latex for the text and Post Script for the figures with a little help from Mathematica. I hope to use this technology to keep the text as error free and current as possible. Special thanks go to my daughter Susan Junghans for an outstanding job of producing Latex files of the entire first edition.

I have been pleased that the first edition was successfully used in high schools, community colleges, regional campuses, and here at College Park. My primary goal for the second edition has been to make it more flexible and current without undermining the foundation. Although many of the changes are small, I think they will add up to a significant improvement of the original. It should now be easier to vary the emphasis on topics and/or supplement aspects of the material. The possibilities for enriching the subject and bringing it to life for the students are limitless. Above all else I hope this Second Edition will be used creatively by instructors at a variety of educational institutions.

College Park, Maryland N.G.M.
October 1994

Contents

Chapter 1

Probability

Life is full of uncertainties. Will it rain tomorrow? Who will win the election? Will interest rates go down next month? Not surprisingly our everyday language abounds with statements indicating some degree of uncertainty with varying amounts of precision. The word "probably" is one of the overworked words in the English language and its meaning depends somewhat on the speaker. A sentence such as, "I will probably see you at lunch time tomorrow," might mean it is almost certain or it is rather unlikely that I will see you at lunch time. The statement, "You have a better chance of winning the tennis match than your opponent" is only slightly more precise. Although it does clearly favor your winning the match, it does not tell us if you have a slight edge or a tremendous advantage. When someone says, "I am ninety per cent certain I will get an A on my English exam tomorrow", he or she is trying to accurately assess the chances or probability in an uncertain situation. Replacing vague statements about uncertain events with numerical measurements of their likelihood is the theme of the mathematical study of probability.

Knowing the probability of an uncertain event does not remove the uncertainty, but it does provide a basis for making rational decisions. For example, if a company knows the probability of landing a particular contract is only one chance in ten, it might decide the time and money needed to prepare a bid are not worth the risk. If the chance of getting the contract was nine chances out of ten, it would almost certainly submit a bid, even though it knew full well they might not win the contract.

Probability as a measurement of likelihood is perhaps easiest understood in the context of games of chance using dice, cards, or spinners. It also occurs very naturally in other areas like genetics, product reliability, decision making, and public opinion polling. It is particularly important to point out that probability is the foundation for statistics and its broad spectrum of applications in our modern world.

The main purpose of this chapter is to introduce mathematical probability in a unified and consistent way (Section 4). This requires some familiarity with sets (Section 1) and with the concept of a sample space (Section 3). The secondary goal of this chapter is to connect the general mathematical theory of probability with the more intuitive classical notion of probability (Section 5) which is based on counting (Section 2).

1.1 The Language of Sets

The language of sets provides a convenient framework for discussing many mathematical problems. A *set* is simply a collection of objects for which membership is completely specified. For example, we cannot construct a set by only specifying that it contains 50 numbers because then it is impossible to determine whether or not the number 10 is in the set. However, the odd numbers larger than 0 and less than 100 is a set because it is easily determined whether or not a specific number is in the set.

We will use capital letters to denote sets and lower case letters to denote objects or elements of a set. If A is a set, then the symbols $x \in A$ are read "x is an element of A" and mean that x is one of the objects in A. Similarly, the symbols $x \notin A$ are read "x is not an element of A" and mean x is an object which is not in A. A set may contain either a finite number or an infinite number of objects or elements.

EXAMPLE 1. The odd numbers between 10 and 20 form a finite set. Denote it by A. Since it is easy to make a list of the elements in A, the most convenient mathematical notation for this set is

$$A = \{11, 13, 15, 17, 19\}.$$

EXAMPLE 2. If we try to use the same method of notation for the set B of odd integers larger than 0 and less than 100, it becomes awkward even though the set is finite. We modify the notation as follows:

$$B = \{1, 3, 5, 7, \ldots, 97, 99\}$$

with the dots denoting omissions in a long list.

EXAMPLE 3. Next let C be the positive integers which are multiples of 3. This is an infinite set. We can write either

$$C = \{3, 6, 9, 12, \ldots\}$$

or

$$C = \{x : x \text{ is a positive integral multiple of } 3\ \}.$$

The right hand side of the second expression is read "the set of all x such that x is a positive integral multiple of 3".

EXAMPLE 4. Finally, let D be the set of all real numbers greater than 1 and less than 10. Here there is no hope of making any sort of list and we can only write

$$D = \{x : 1 < x < 10\}.$$

because this set contains an infinite host of numbers like $\sqrt{2}$, $\sqrt[3]{5}$, π, etc..

Let A and B be two sets. If every element of A is also an element of B we say A is contained in B and we write $A \subset B$. If $A \subset B$ and $B \subset A$, then $A = B$.

EXAMPLE 5. Let $A = \{2,3,4\}$, $B = \{1,2,3,4,5,6\}$, and $C = \{2,4,6)$. Then $A \subset B$ because 2, 3, and 4 are also elements of B. However, even though 2 and 4 are in C, A is not contained in C because not every element of A is also an element of C. Specifically, $3 \in A$ and $3 \notin C$.

The set which contains no elements is called the *null or empty set* and is denoted by $\emptyset$. At the other extreme the collection of all elements being considered in a given situation is called the *universal set* and will be denoted by U.

There are three basic operations that are used to build new sets from given ones. We can consider the elements outside a set, we can combine the elements from a number of sets to form a large set, and we can select the elements common to a number of sets. By using several of these operations together a great variety of sets can be constructed from a few given sets.

To define these operations, let A and B be subsets of the universal set U. First the *complement* of A is defined by

$$A' = \{x : x \notin A\}.$$

Notice that the definition gives the notation A' for the complement of A as well as stating that the complement of A consists of the points in

U which are not in A. Next the *union* of A and B is defined by

$$A \cup B = \{x : x \in A \text{ or } x \in B\}.$$

The "or" in this definition does not exclude the possibility that x is in both A and B. Finally, the *intersection* is defined by

$$A \cap B = \{x : x \in A \text{ and } x \in B\}.$$

Sample Problem 1. *Let* $U = \{1,2,3,4,5,6,7,8,9,10\}$, $A = \{2,4,6\}$, $B = \{1,2,4,6\}$, $C = \{2,4,6,8,10\}$, $D = \{1,3,5\}$ *and* $E = \{4,5,6\}$. *List the elements in the following sets:*

(a) B', *(d)* $A \cap E$,
(b) $A \cup D$, *(e)* $A \cap D$,
(c) $D \cup E$, *(f)* $(A \cup E)' \cap (B \cup C')$.

SOLUTION: The list of elements in each of these sets will be made using the definitions of complement, union, and intersection.

(a) For B' we must list the elements in U which are not in B, that is

$$B' = \{3,5,7,8,9,10\}.$$

(b) To list the elements in $A \cup D$ we simply combine the list of elements in A with the list of elements in D to get

$$A \cup D = \{2,4,6,1,3,5\} = \{1,2,3,4,5,6\}.$$

We rewrote the list in a more natural order, but the elements in a set may be listed in any order.

(c) For $D \cup E$ we must also combine the lists of elements in two sets. Although 5 appears in the list of elements for both D and E, we do not write it down twice when listing the elements in $D \cup E$. Thus,

$$D \cup E = \{1,3,5,4,6\} = \{1,3,4,5,6\}.$$

(d) To list the elements in $A \cap E$ we systematically work through the list of elements for A, writing down only those which are also elements of E as shown.

$$\begin{array}{llll} A = \{ & 2, & 4, & & 6\,\} \\ & & \updownarrow & & \updownarrow \\ E = \{ & & 4, & 5, & 6\,\} \end{array}$$

In particular, we do not write down 2 because it is not in E, but we do write down 4 and 6 because they are in E. Thus,

$$A \cap E = \{4, 6\}.$$

(e) When we try to carry this out for $A \cap D$, we can not find any elements in A which are also in D. This means that

$$A \cap D = \emptyset.$$

(f) Finally, to determine the elements in $(A \cup E)' \cap (B \cup C')$, we must start with the simpler parts, $A \cup E$ and C'; then do the next level, $(A \cup E)'$ and $B \cup C'$; and finally intersect $(A \cup E)'$ and $B \cup C'$ to get the answer. Notice that $(A \cup E)'$ means the complement of the union of A and E while $(B \cup C')$ means the union of B and complement of C. From the lists of elements in the sets A, C, and E we see that $A \cup E = \{2, 4, 5, 6\}$ and $C' = \{1, 3, 7, 9\}$. From these lists it follows that

$$(A \cup E)' = \{1, 3, 7, 8, 9, 10\}$$

and

$$(B \cup C') = \{1, 2, 3, 4, 5, 6, 7, 9\}.$$

Finally by writing down the elements in both the lists for $(A \cup E)'$ and for $(B \cup C')$, we get

$$(A \cup E)' \cap (B \cup C') = \{1, 3, 7, 9\}$$

This completes the solution of a sample problem which will always be indicated by the following symbol: $\square$

Although the definitions of union and intersection were presented for two sets, these concepts make perfect sense for any number of sets. Specifically, the union of any number of sets consists of all the elements which belong to at least one of the sets, and the intersection of any number of sets consists of all the elements which are common to all the sets.

EXAMPLE 6. Using the sets $A = \{2, 3, 4\}$, $B = \{1, 2, 3, 4, 5, 6\}$, and $C = \{2, 4, \}$ from Example 5 to illustrate unions and intersections of three sets, we have

$$A \cup B \cup C = \{1, 2, 4, 6, 8, 10\}$$

and

$$B \cap C \cap E = \{4, 6\}.$$

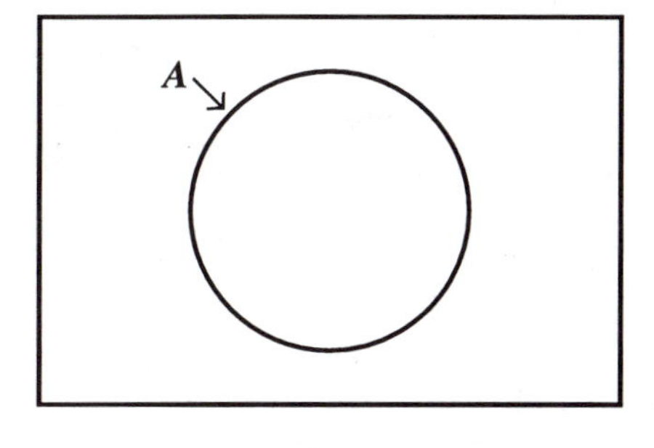

Figure 1

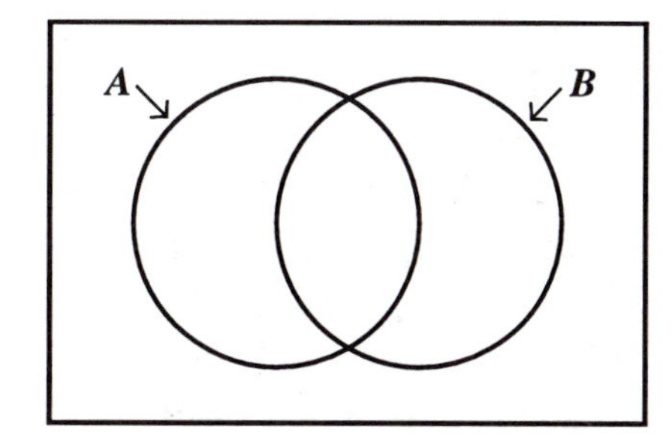

Figure 2

Thus far we have been working with sets primarily by making lists, which is fine when you have a finite list of the elements in U. We would also like to have a way of picturing complements, unions, and intersections of sets to use when we don't have a list of the elements in U. The *Venn diagram* is just such a pictorial device and is often useful for reasoning through problems involving two or three sets. A Venn diagram consists of a rectangle which represents the universal set U and circles inside the rectangle which represent specific subsets of U. For example, Figure 1 is a Venn diagram for one subset A of U. The inside of the circle represents A and the region between the circle and the rectangle represents A'. Of course, to make a Venn diagram showing two sets, A and B, we use two circles, but unless we are absolutely sure that $A \cap B = \emptyset$ these circles should be drawn so that they overlap as in Figure 2.

The next Sample Problem shows how Venn diagrams can be used to determine general relationships between two or three sets.

Sample Problem 2. *Is the formula*

$$(A \cup B)' = A' \cap B'$$

always true?

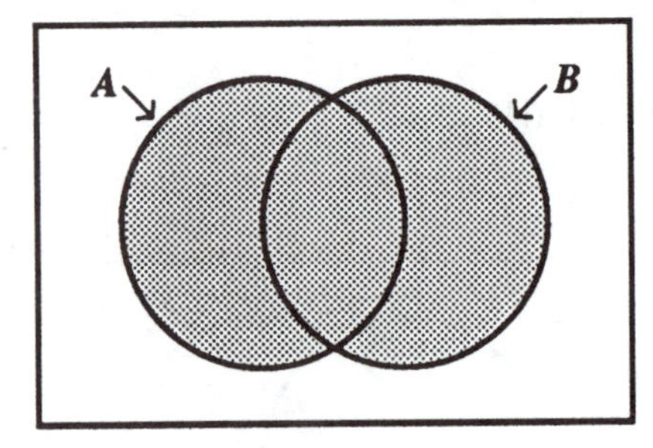

Figure 3

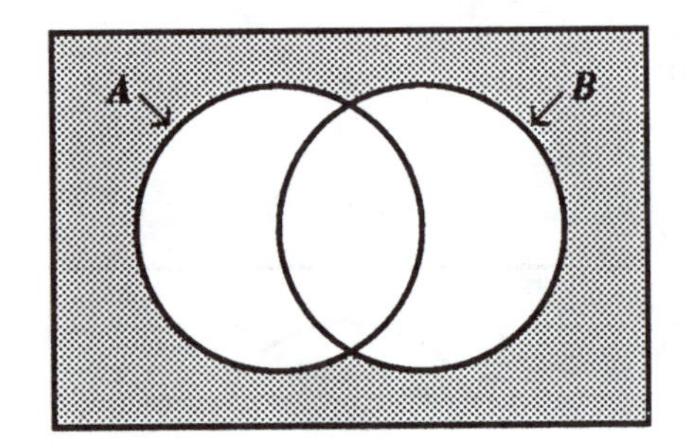

Figure 4

SOLUTION: For each side of the equation we will construct a Venn diagram with the set on that side of the equation shaded. If the same region is shaded in both constructions, the equation is always true. If different regions are shaded in the two diagrams, the equation is not true.

Figure 3 is a Venn diagram with $A \cup B$ shaded To shade the complement we just switch the shaded and unshaded parts as shown in Figure 4.

Figures 5 and 6 show A' and B' respectively shaded. To identify the intersection of these two sets we shade the portion of a Venn diagram that is shaded in both Figure 5 and 6 because the intersection consists of the points common to both sets. Doing this leads right back to Figure 4 and the conclusion that $(A \cup B)' = A' \cap B'$ because both constructions produce the same Venn diagram. □

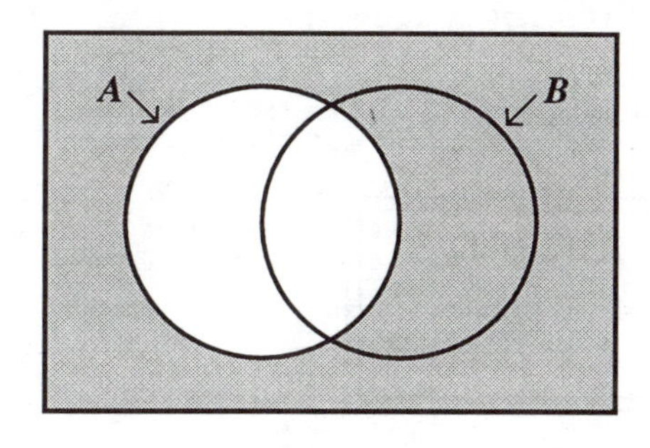

Figure 5

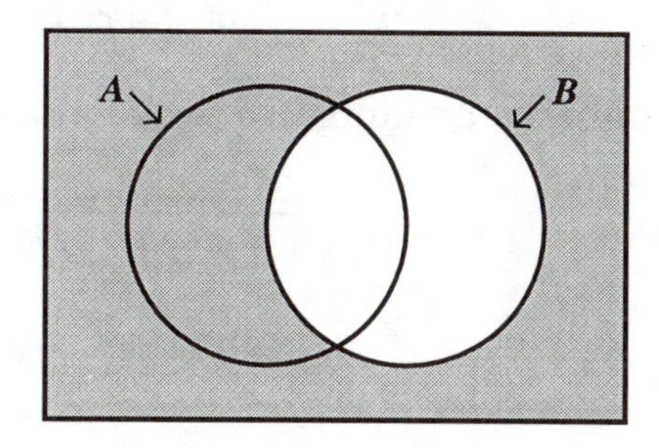

Figure 6

A Venn diagram for two sets, A and B, divides the rectangle into the four sets $A \cap B$, $A \cap B'$, $A' \cap B$, and $A' \cap B'$. These four parts of a Venn diagram for two sets will be called *cells*. These four cells are labeled in Figure 7. We know from the Sample Problem 2 that the outer cell, $(A \cup B)'$, is the same as $A' \cap B'$ which is the label used for it. These cells will be used in the next section to solve counting problems.

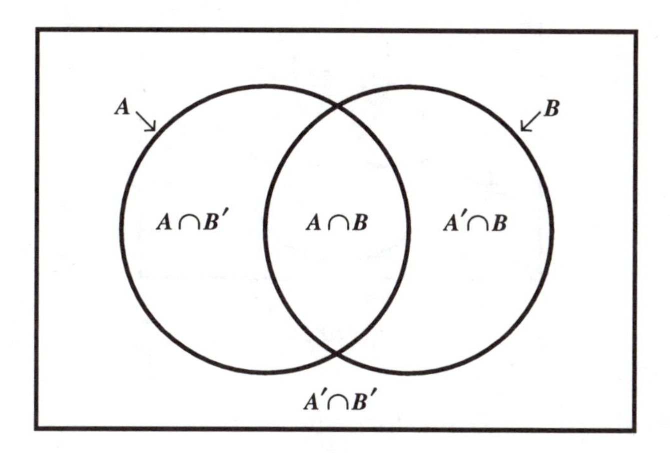

Figure 7

The formula

$$(A \cup B)' = A' \cap B'$$

from Sample Problem 2 is one of de Morgan's laws for two sets. The other one is

$$(A \cap B)' = A' \cup B'.$$

Similarly, de Morgan's laws for three sets are:

$$(A \cup B \cup C)' = A' \cap B' \cap C'$$

$$(A \cap B \cap C)' = A' \cup B' \cup C'$$

A Venn diagram for three sets - A, B, and C - is illustrated in Figure 8. Notice it divides the inside of the rectangle into eight non-overlapping cells. The shaded region inside all three circles is clearly $A \cap B \cap C$ and the region outside of all three circles is $(A \cup B \cup C)' = A' \cap B' \cap C'$ by de Morgan's laws. The exercises at the end of this section will begin exploring the other eight cells in preparation for the next section.

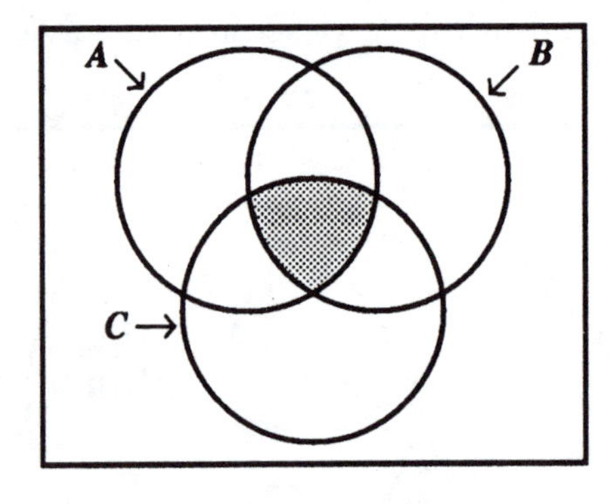

Figure 8

Sample Problem 3. *Is the formula*

$$(A \cup B) \cap C' = (A \cap B) \cup C'$$

always true?

SOLUTION: We use the same procedure as in Sample Problem 2. In Figure 9 and 10 $(A \cup B)$ and C' are shaded respectively. To construct a

Venn diagram with $(A \cup B) \cap C'$ shaded, we must shade those portions shaded for both $A \cup B$ (Figure 9) and C' (Figure 10). This is done in Figure 13.

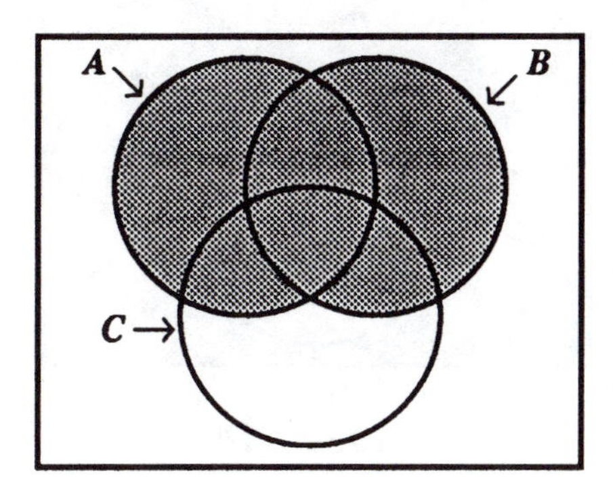

Figure 9

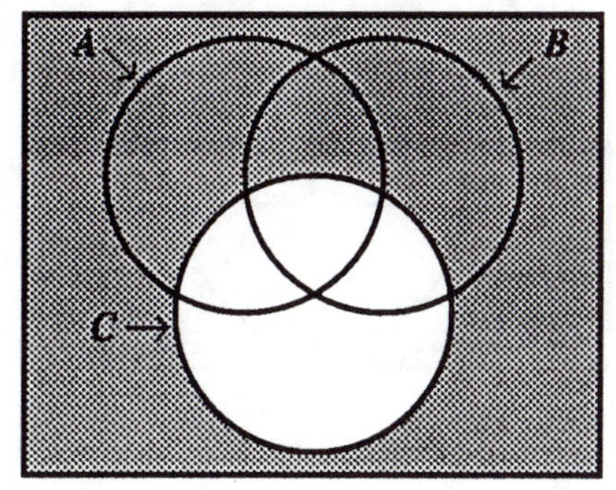

Figure 10

Turning to the set on the right side of the equation, Figures 11 and 12 show $A \cap B$ and C' shaded. Now to construct a Venn diagram with $(A \cap B) \cup C'$ shaded, we must shade those portions shaded for either $A \cap B$ (Figure 11) or C' (Figure 12). This is done in Figure 14.

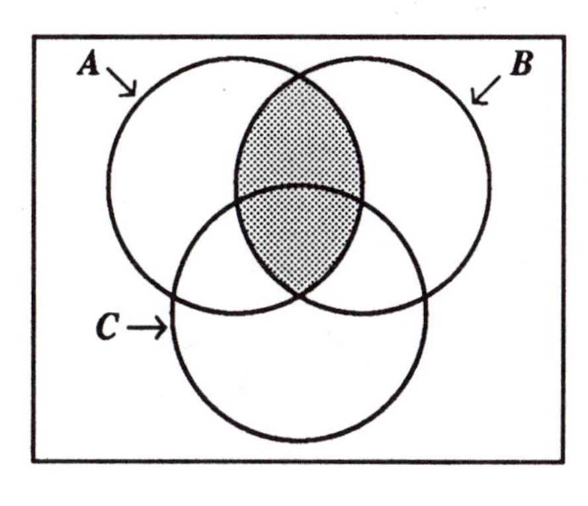

Figure 11

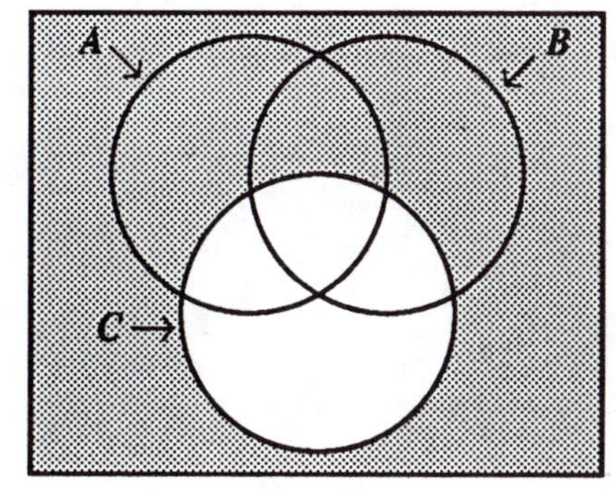

Figure 12

The final step is to compare the shaded regions in Figures 13 and 14. Since the shaded regions are different, it is not true that $(A \cup B) \cap C' = (A \cap B) \cup C'$ and this is not a valid formula for sets. □

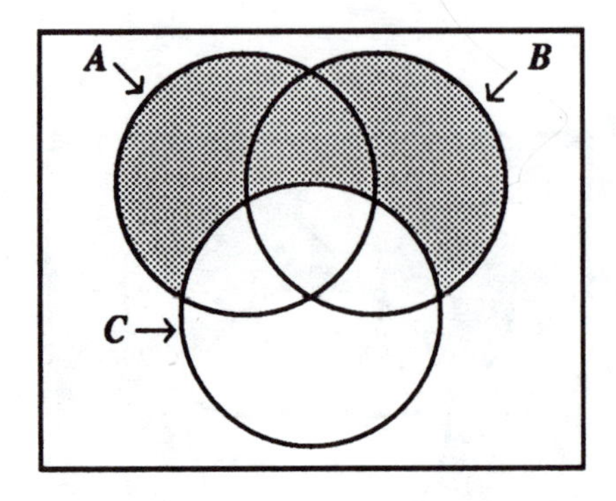

Figure 13

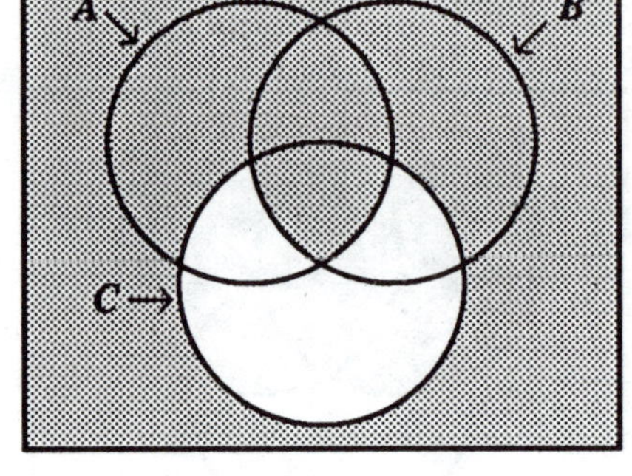

Figure 14

Exercises

1. Consider the following sets of integers: $A = \{2,4,8,16\}$, $B = \{2,4,8,10,14,16\}$, and $C = \{2,8,10,16\}$.

   (a) Is either $A \subset B$ or $B \subset A$ true?

   (b) Is either $A \subset C$ or $C \subset A$ true?

   (c) Is either $B \subset C$ or $C \subset B$ true?

2. Let D, E, and F be the subsets of the alphabet given by $D = \{c,f,i,l,o,u,x\}$, $E = \{c,f,l,m,r,s,u\}$, and $F = \{f,l,r\}$.

   (a) Is either $D \subset E$ or $E \subset D$ true?

   (b) Is either $E \subset F$ or $F \subset E$ true?

   (c) Is either $D \subset F$ or $F \subset D$ true?

3. Let $U = \{1,2,3,4,5,6,7,8\}$, $A = \{4,5,6,7,8\}$, $B = \{3,5,8\}$, $C = \{4,5\}$, and $D = \{2,3,4,5,6\}$. List the elements in the following sets:

   (a) $B \cup C$,

   (b) $B \cap D$,

   (c) A',

   (d) $(A \cap B') \cap (C \cup D)'$.

4. Let $U = \{a,b,c,d,e,f,g,h\}$, $A = \{a,b,c\}$, $B = \{b,d,f,h\}$, and $C = \{f,g,h\}$. List the elements in the following sets:

   (a) $A \cup B$,

   (b) $B \cap C$,

   (c) B',

   (d) $[(A \cap B') \cup C]'$.

5. Let the universal set be $U = \{0,1,2,3,4,5,6\}$. Let $A = \{0,2,4,6\}$, $B = \{0,1,2,4,5,6\}$, and $C = \{3,4,5\}$. Find the following sets:

   (a) $B \cap C$,

(b) $A' \cup C$,

(c) $(A \cup B')'$.

6. Let U = { Bill, Carol, Ellen, John, Sue, Tom } be the universal set and consider the following subsets:

$$\begin{aligned} A &= \{ \text{ Bill, Ellen, John, Sue } \}, \\ B &= \{ \text{ Carol, Ellen, John, Sue } \}, \\ C &= \{ \text{ Ellen, John, Sue, Tom } \}. \end{aligned}$$

Find $A' \cup B'$, $(A \cap B')'$, and $B \cup C'$.

7. Let $U = \{1, 2, 3, 4, 5, 6, 7, 8, 9, 10\}$, $A = \{1, 3, 4, 5, 9\}$, $B = \{2, 3, 4, 9, 10\}$, $C = \{5, 6, 7\}$, and $D = \{1, 8, 10\}$. Find the following sets:

(a) $A \cup C$,

(b) $A \cap B$,

(c) A',

(d) $(C' \cap D') \cap (A \cup C)$.

8. Consider the subsets $P = \{a, c, e\}$, $Q = \{a, b, c\}$, and $R = \{c, d, e\}$ of the universal set $U = \{a, b, c, d, e, f, g\}$. List the elements in each of the following sets:

(a) $(P \cap Q)$,

(b) $(P \cup Q \cup R)'$,

(c) $P' \cup Q'$,

(d) $P' \cap Q' \cap R'$.

9. Let the universe be $T = \{1, 3, 5, 7, 9, 11, 13\}$, and let $D = \{1, 7, 9, 13\}$, $E = \{3, 9, 11, 13\}$, and $F = \{5, 7, 11, 13\}$. Determine the elements in the following sets:

(a) $D' \cap (E \cup F)$,

(b) $(D \cup E)' \cup (D \cap E)$,

(c) $(F' \cap E) \cup D$.

10. For each of the following make a Venn diagram showing three sets labeled A, B, C with the given set shaded:

    (a) $A' \cap B' \cap C$,

    (b) $A \cap B \cap C'$,

    (c) $(A \cup B) \cap C'$,

    (d) $(A' \cup B) \cap C$.

11. Make and label a Venn diagram for three sets - A, B, and C. Then shade the region corresponding to $(A \cup C) \cap (B \cup C)$.

12. Shade the region $B \cap (A \cap C)'$ in a Venn diagram for the sets A, B, and C.

13. Make two Venn diagrams showing 3 sets clearly labeled A, B, and C.

    (a) In the first one shade in $(A \cap C) \cup B$.

    (b) In the second one shade in $A \cap (B \cup C)$.

    (c) Is it always true that $(A \cap C) \cup B = A \cap (B \cup C)$?

14. Determine all the subsets of the universal set $U = \{2, 4, 6, 8\}$.

15. Make a Venn diagram for three sets labeled P, Q, and R and shade the set $(P \cap Q') \cup (P \cap R)$. Do the same for

$$(P \cap Q' \cap R') \cup (P' \cap Q \cap R).$$

16. Use Venn diagrams to verify deMorgan's second law,

$$(A \cap B \cap C)' = A' \cup B' \cup C'$$

17. Let A, B, and C be subsets of U. Is it always true that

$$(A \cap B) \cup C = (A \cup B) \cap C?$$

1.2 The Rudiments of Counting

If A is a finite set, the elements in A can be counted and the number of elements in A will be denoted by $\mathbf{n}(A)$. What are the obvious rules which the counting function $\mathbf{n}(A)$ obeys? Clearly $\mathbf{n}(A)$ is an integer which cannot be negative and $\mathbf{n}(\emptyset) = 0$. But the most important fact is

$$\mathbf{n}(A \cup B) = \mathbf{n}(A) + \mathbf{n}(B) \text{ when } A \cap B = \emptyset.$$

In other words, when A and B have nothing in common, we can count the number of objects in A and the number of objects in B and then add to get the number of objects in $A \cup B$.

When $A \cap B = \emptyset$ we say A and B are *disjoint*. Let C be a finite set. If we can partition C into two disjoint sets, that is, $C = A \cup B$ and $A \cap B = \emptyset$, then $\mathbf{n}(C) = \mathbf{n}(A) + \mathbf{n}(B)$. This works just as well with more than two sets and suggests the following definition: $A_1, A_2, A_3, \ldots, A_n$ is a *partition* of C provided

$$C = A_1 \cup A_2 \cup A_3 \cup \ldots \cup A_n$$

and

$$A_i \cap A_j = \emptyset \text{ when } i \neq j.$$

The expression "$A_i \cap A_j = \emptyset$ when $i \neq j$" means that every pair of sets from $A_1, A_2, A_3, \ldots, A_n$ is disjoint. If $A_i \cap A_j = \emptyset$ when $i \neq j$, we say $A_1, A_2, A_3, \ldots, A_n$ are *mutually exclusive*; and if

$$C = A_1 \cup A_2 \cup A_3 \cup \ldots \cup A_n,$$

we say $A_1, A_2, A_3, \ldots, A_n$ *exhaust* C. Thus a partition of C is simply a collection of mutually exclusive sets which exhaust C.

We can now state the first fundamental principle of counting, which is just a fancy version of $\mathbf{n}(A \cup B) = \mathbf{n}(A) + \mathbf{n}(B)$ when $A \cap B = \emptyset$.

PARTITION PRINCIPLE

If $A_1, A_2, A_3, \ldots, A_n$ is a partition of a finite set C, then

$$\mathbf{n}(C) = \mathbf{n}(A_1) + \mathbf{n}(A_2) + \mathbf{n}(A_3) + \ldots + \mathbf{n}(A_n).$$

EXAMPLE 1. Let U be the set of students enrolled in the University of Maryland, let A_1 be the subset of students born in January, let A_2 be the subset of students born in February, etc.. Then $A_1, A_2, A_3, \ldots, A_{12}$ is a partition of U because every student was born in some month but no student has a birthday in two different months. In this example the Partition Principle says that you can determine the total number of students by determining the number born in each of the twelve months and adding.

Suppose the universal set U is finite and A is a subset of U. Then A and A' form a partition of U and

$$\mathrm{n}(U) = \mathrm{n}(A) + \mathrm{n}(A')$$

or

$$\mathrm{n}(U) - \mathrm{n}(A) = \mathrm{n}(A').$$

Now let B be another subset of U. Is there a formula for computing $\mathrm{n}(A \cup B)$ when $A \cap B \neq \emptyset$? A Venn diagram (Figure 1) in which each cell is labeled will make the ideas easier to follow. From Figure 1 it is clear that the four cells of the Venn diagram are a partition of U, and in particular $A \cap B$ and $A \cap B'$ form a partition of A. Thus

$$\mathrm{n}(A) = \mathrm{n}(A \cap B) + \mathrm{n}(A \cap B')$$

or

$$\mathrm{n}(A) - \mathrm{n}(A \cap B) = \mathrm{n}(A \cap B').$$

Similarly

$$\mathrm{n}(B) - \mathrm{n}(A \cap B) = \mathrm{n}(A' \cap B).$$

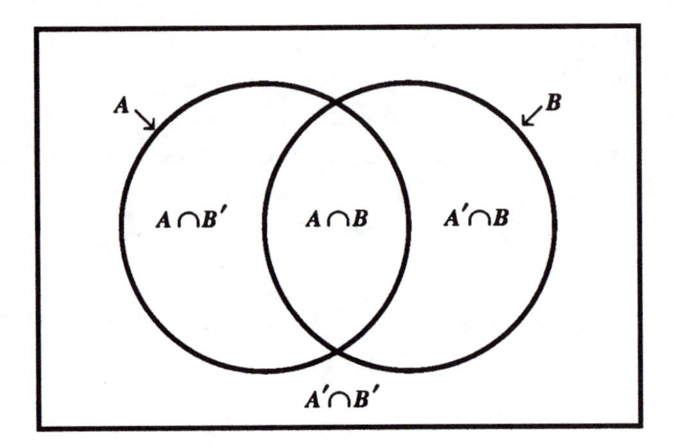

Figure 1

Next observe in Figure 1 that $A \cap B$, $A \cap B'$ and $A' \cap B$ is a partition of $A \cup B$ and hence

$$\mathbf{n}(A \cup B) = \mathbf{n}(A \cap B) + \mathbf{n}(A \cap B') + \mathbf{n}(A' \cap B).$$

By the two previous formulas we can substitute $\mathbf{n}(A) - \mathbf{n}(A \cap B)$ for $\mathbf{n}(A \cap B')$ and $\mathbf{n}(B) - \mathbf{n}(A \cap B)$ for $\mathbf{n}(A' \cap B)$ and get

$$\mathbf{n}(A \cup B) = \mathbf{n}(A \cap B) + \mathbf{n}(A) - \mathbf{n}(A \cap B) + \mathbf{n}(B) - \mathbf{n}(A \cap B)$$

or

$$\mathbf{n}(A \cup B) = \mathbf{n}(A) + \mathbf{n}(B) - \mathbf{n}(A \cap B).$$

A less formal way to understand this fundamental formula is the following: If we try to find $\mathbf{n}(A \cup B)$ by counting all the elements in A and then all the elements in B, everything in $A \cap B$ is counted twice. Thus to obtain $\mathbf{n}(A \cup B)$ we must subtract our error, which is precisely $\mathbf{n}(A \cap B)$, from $\mathbf{n}(A) + \mathbf{n}(B)$.

To summarize, if A and B are subsets of a finite universal set U, then the following formulas hold:

$$\mathbf{n}(A') = \mathbf{n}(U) - \mathbf{n}(A)$$

$$\mathbf{n}(A \cap B') = \mathbf{n}(A) - \mathbf{n}(A \cap B)$$

$$\mathbf{n}(A \cup B) = \mathbf{n}(A) + \mathbf{n}(B) - \mathbf{n}(A \cap B).$$

Sample Problem 1. *Let A and B be subsets of U. If $\mathbf{n}(U) = 100$, $\mathbf{n}(A) = 55$, $\mathbf{n}(B) = 50$, and $\mathbf{n}(A \cap B) = 20$, how many elements are in the following sets:*

(a) A', *(b)* $A \cap B'$,
(c) $A \cup B$, *(d)* $A' \cap B'$?

SOLUTION: These questions can be answered using the above formulae directly or indirectly with a Venn diagram. Both approaches will be presented, starting with the direct application of the formulae:

(a) $\mathbf{n}(A') = \mathbf{n}(U) - \mathbf{n}(A) = 100 - 55 = 45$,

(b) $\mathbf{n}(A \cap B') = \mathbf{n}(A) - \mathbf{n}(A \cap B) = 55 - 20 = 35$,

(c) $\mathbf{n}(A \cup B) = \mathbf{n}(A) + \mathbf{n}(B) - \mathbf{n}(A \cap B) = 55 + 50 - 20 = 85$,

(d) and finally by de Morgan's law $A' \cap B' = (A \cup B)'$ and hence

$$\mathbf{n}(A' \cap B') = \mathbf{n}(U) - \mathbf{n}(A \cup B) = 100 - 85 = 15.$$

This completes the first solution.

Alternatively we can make a Venn diagram for two sets and in each of the four cells write the number of elements contained in that cell. First we write 20 in the center cell because $\mathbf{n}(A \cap B) = 20$ is given. Since 20 of the 55 elements in A are also in B, the remaining 35 must be in $A \cap B'$ as shown in Figure 2. Similarly, there are 30 elements in $A' \cap B$. We have now accounted for $20 + 35 + 30 = 85$ of the elements in U, hence there must be 15 in the remaining outer cell to complete

Figure 2.

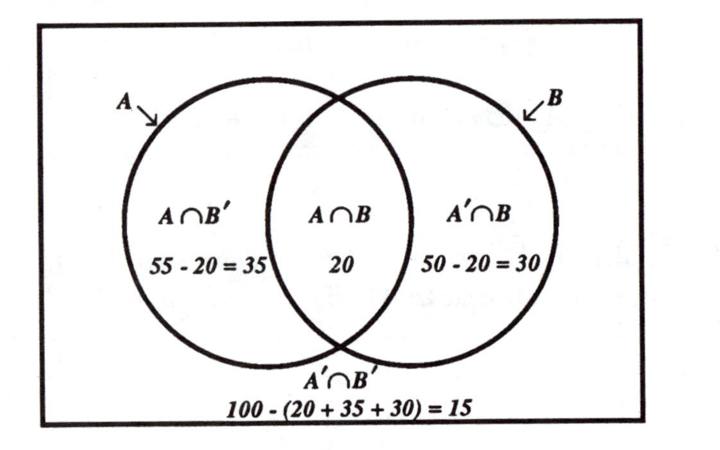

Figure 2

To answer the original questions from the Sample Problem we read the answers off the Venn diagram as follows:

(a) Adding the numbers in the two cells outside of A, yields

$$\mathbf{n}(A') = 30 + 15 = 45.$$

(b) Since $A \cap B'$ is one of the cells,

$$\mathbf{n}(A \cap B') = 35.$$

(c) The set $A \cup B$ consists of the thre cells lying inside the circles, hence

$$\mathbf{n}(A \cup B) = 35 + 20 + 30 = 85.$$

(d) Lastly $A \cap B$ is the outer cell by de Morgan and

$$\mathbf{n}(A' \cap B') = 15.$$

Although we did not write down any formulas in the second solution, we still used the basic counting principles implicitly in our reasoning. □

The Venn diagram approach is particularly effective when dealing with problems for three sets similar to Sample Problem 1. To use a Venn diagram to solve such problems it is necessary to identify and label the eight cells in a Venn diagram for three sets. The cell in all three circles appearing in the center of the diagram is obviously $A \cap B \cap C$ as shown in Figure 5. The one around the outside is the complement of $A \cup B \cup C$ and by de Morgan

$$(A \cup B \cup C)' = A' \cap B' \cap C'.$$

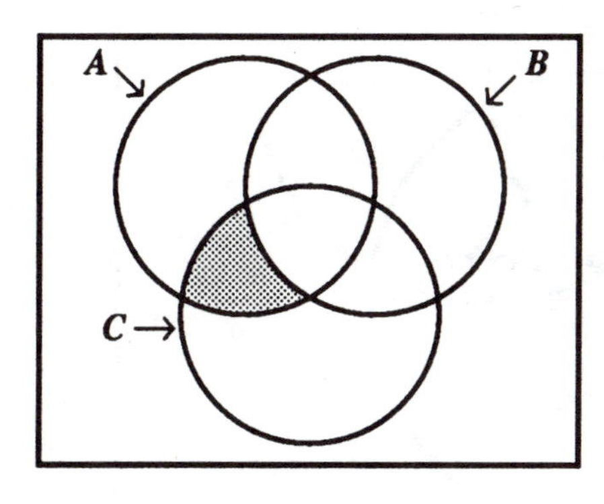

Figure 3

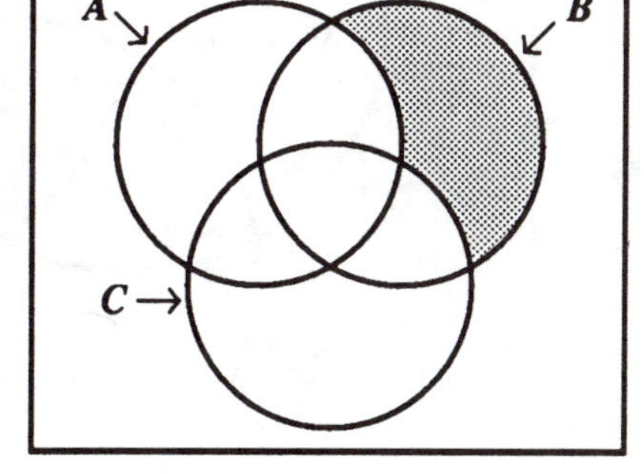

Figure 4

Next consider the cell shaded in Figure 3. It lies inside A and C but outside B. Hence it must be $A \cap B' \cap C$. The other two similarly shaped cells are $A \cap B \cap C'$ and $A' \cap B \cap C$. The cell shaded in Figure 4 is only contained in B lying outside A and C. Thus it is $A' \cap B \cap C'$. The last two cells are likewise only in one circle and are $A \cap B' \cap C'$ and $A' \cap B' \cap C$. This completes the labeling of Figure 5 which will be helpful in following the solution of the next problem.

Sample Problem 2. *Suppose A, B, and C are three subsets of U and the following data are given:* $\mathbf{n}(U) = 150$, $\mathbf{n}(A) = 35$, $\mathbf{n}(B) = 40$, $\mathbf{n}(C) = 45$, $\mathbf{n}(A \cap B) = 10$, $\mathbf{n}(A \cap C) = 7$, $\mathbf{n}(B \cap C) = 6$, *and* $\mathbf{n}(A \cap B \cap C) = 2$. *Determine the number of elements in each of the following sets:*

(a) $A \cap B \cap C'$,

(b) $A \cap B' \cap C'$,

(c) $A \cup B \cup C$,

(d) $A' \cap B' \cap C'$.

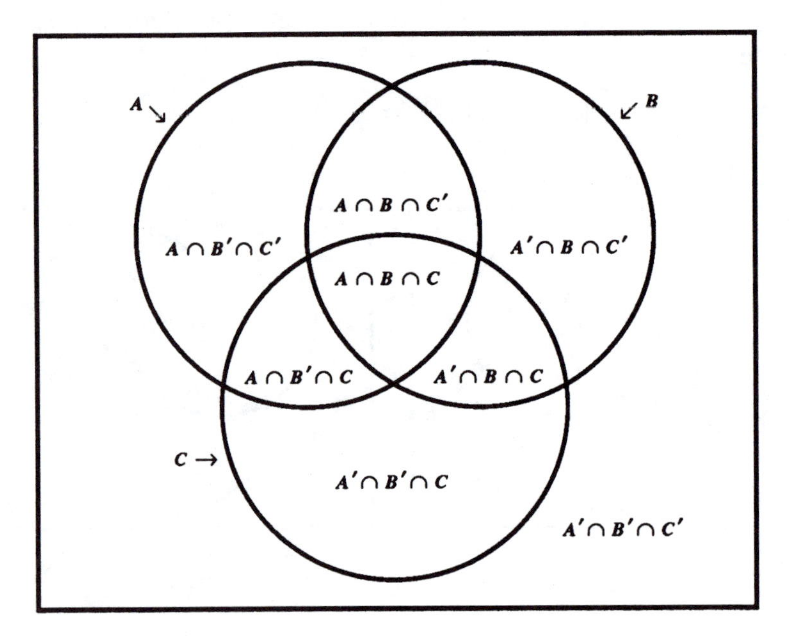

Figure 5

SOLUTION: We will use the same method that we employed in the second solution of Sample Problem 1 except that here the Venn diagram is partitioned into eight cells instead of just four. The first step is to find in the given data the number of elements in at least one of the cells of the Venn diagram. In this problem it is $\mathbf{n}(A \cap B \cap C) = 2$ which is the center cell. After entering this number in the Venn diagram (Figure 6), we look for neighboring cells whose elements can now be counted. Since there are 10 elements in $A \cap B$ and 2 of them are in $A \cap B \cap C$, there must be 8 in $A \cap B \cap C'$ which is the other cell in $A \cap B$. Thus

$$\mathbf{n}(A \cap B \cap C') = 8.$$

Similarly

$$\mathbf{n}(A \cap B' \cap C) = 5$$

and

$$\mathbf{n}(A' \cap B \cap C) = 4.$$

Our progress thus far is shown in Figure 6.

To find the number of elements in the cell $A \cap B' \cap C'$ notice that A is partitioned into four pieces, 3 of them already have numbers in them, and the fourth is $A \cap B' \cap C'$. Because there are 35 objects in A we must have

$$5 + 2 + 8 + \mathbf{n}(A \cap B' \cap C') = 35$$

and then by subtracting

$$\mathbf{n}(A \cap B' \cap C') = 20.$$

Similarly

$$8 + 2 + 4 + \mathbf{n}(A' \cap B \cap C') = 40$$

implies that

$$\mathbf{n}(A' \cap B \cap C') = 26$$

and

$$5 + 2 + 4 + \mathbf{n}(A' \cap B' \cap C) = 45$$

implies that

$$\mathbf{n}(A' \cap B' \cap C) = 34.$$

These numbers are now entered in the Venn diagram as shown in Figure 7.

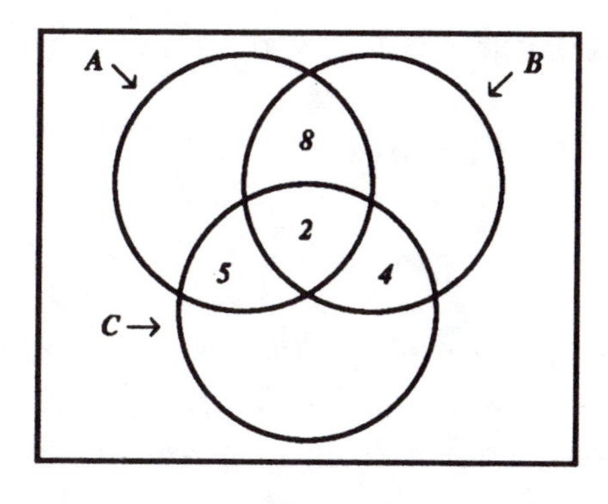

Figure 6

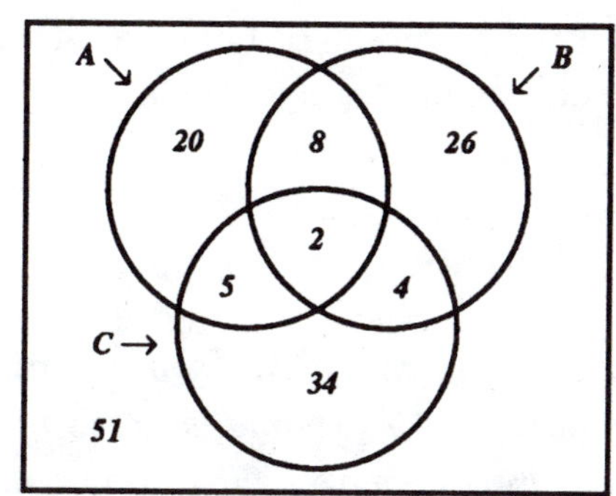

Figure 7

We have now calculated the number of elements in seven of the eight cells. Since the sum of the numbers of the elements in all eight cells must equal the total number of elements in U, the number of elements in the last cell can be found by subtracting the sum of the numbers in the first seven from the total number in U. Specifically

$$\mathbf{n}(A' \cap B' \cap C') = 150 - (2 + 8 + 5 + 4 + 20 + 26 + 34) = 51.$$

The completed Venn diagram is shown in Figure 7 and with it we can easily finish the solution.

(a) The first set, $A \cap B \cap C'$, is one of the cells in the Venn diagram, and we have already determined the number of elements in it, namely

$$\mathbf{n}(A \cap B \cap C') = 8$$

(b) The second set is also one of the cells in the Figure 7 and

$$\mathbf{n}(A \cap B' \cap C') = 20.$$

(c) We also have the number of elements in $A \cup B \cup C$ at our fingertips. Because $A \cup B \cup C$ is made up of the seven cells lying inside the three circles

$$\mathbf{n}(A \cup B \cup C) = 2 + 8 + 5 + 4 + 20 + 26 + 34 = 99.$$

(d) Finally, $A' \cap B' \cap C'$ is the cell lying entirely outside all three circles and hence

$$\mathbf{n}(A' \cap B' \cap C') = 51$$

And we are done. □

Sample Problem 3. *In a dormitory containing 500 rooms there are televisions in 175 rooms, stereos in 225 rooms, and refrigerators in 190 rooms. Furthermore, 90 rooms contain both a television and a stereo, 75 contain both a television and a refrigerator, and 85 contain both a stereo and refrigerator. Finally, 40 rooms contain a television and a stereo but no refrigerator.*

(a) *How many rooms do not contain a refrigerator, a stereo, or a television?*

(b) *How many rooms contain exactly one of these three kinds of appliances?*

(c) *How many contain at least two of the three?*

SOLUTION: Let T, S, and R denote respectively the subsets of rooms in the dormitory which contain respectively a television, a stereo, and a refrigerator. The information given in the statement can now be written as follows:

$$\mathbf{n}(T) = 175, \mathbf{n}(S) = 225, \mathbf{n}(R) = 190$$

$$\mathbf{n}(T \cap S) = 90, \mathbf{n}(T \cap R) = 75, \mathbf{n}(S \cap R) = 85$$

$$\mathbf{n}(T \cap S \cap R') = 40.$$

(Of course, U consists of all 500 rooms in the dormitory).

The key cell is $T \cap S \cap R'$ which contains 40 elements. Since $T \cap S$ consists of the two cells $T \cap S \cap R$ and $T \cap S \cap R'$, and since there are 90 elements in $T \cap S$, there must be 50 elements in $T \cap S \cap R$. We have now determined the number of elements in the center cell and the number of elements in the rest of the cells can be determined as in Sample Problem 2. It is a good idea to do these calculations yourself and to compare your results with those in Figure 8 which will be used to answer the specific questions.

(a) This question is the same as asking for the number of elements in $T' \cap S' \cap R'$ and from Figure 8 we see that

$$\mathbf{n}(T' \cap S' \cap R') = 110.$$

(b) There are three possible ways for a room to contain exactly one of these items, namely, containing a television but no stereo or refrigerator; containing a stereo but no television or refrigerato; containing a refrigerator but no television or stereo. These three possibilities correspond to the three cells $T \cap S' \cap R'$, $T' \cap S \cap R'$, and $T' \cap S' \cap R$ and provide a natural partition of the set of rooms containing exactly one of these three items. Thus the answer is

$$\mathbf{n}(T \cap S' \cap R') + \mathbf{n}(T' \cap S \cap R') + \mathbf{n}(T' \cap S' \cap R) =$$

$$60 + 100 + 80 = 240$$

where the numbers are taken from the Venn diagram.

(c) At least two means two or more, so in this problem we must count the number of rooms with exactly two and those with exactly three of these appliances. In Figure 8 the set of rooms containing at least two of these items is shaded. Notice that there is a natural partition of this set into 4 cells, namely $T \cap S \cap R'$, $T \cap S' \cap R$, $T' \cap S \cap R$, and $T \cap S \cap R$. The number of rooms containing at least two of these items is

$$\mathbf{n}(T \cap S \cap R') + \mathbf{n}(T \cap S' \cap R) + \mathbf{n}(T' \cap S \cap R) + \mathbf{n}(T \cap S \cap R) =$$

$$40 + 25 + 35 + 50 = 150.$$

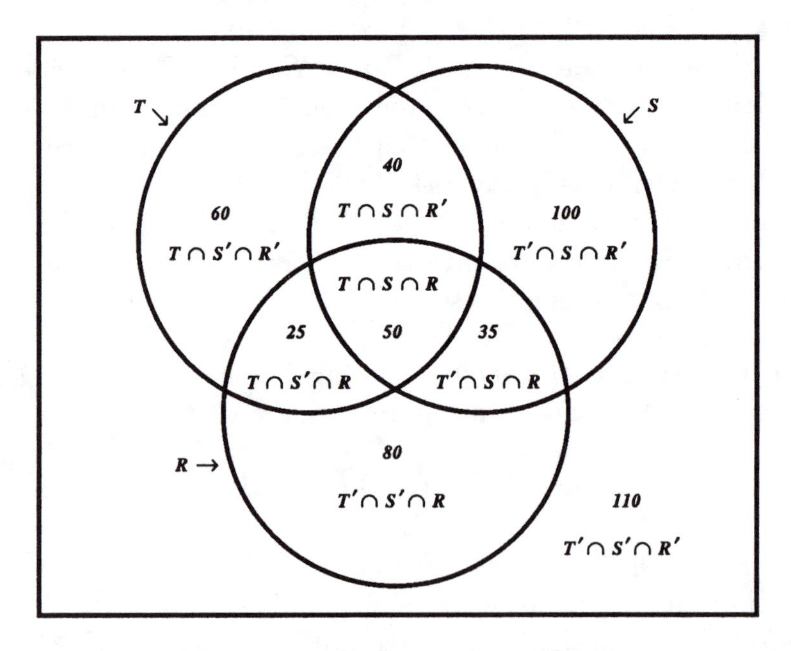

Figure 8

The strategy used here and in similar problems is always to locate a cell with a known number of elements in it and then use the Partition Principle to find the number of elements in near by cells and gradually determine the number of elements in every cell. When the number of elements in every cell is known, then specific questions can be answered by adding the numbers in appropriate cells. □

Although we have not formally invoked the Partition Principle, we have used it repeatedly in these three examples. This principle is just a mathematical statement of the strategy of breaking a complicated set into simple pieces with nothing in common, counting the number of objects in each of the simple pieces, and adding to get the total number of objects in the original set. The same strategy also applies to the calculation of probabilities.

Exercises

1. Let $C = \{A, E, F, H, I, K, L, M, N, T, V, W, X, Y, Z\}$. (Each capital letter in C can be made using two, three, or four straight lines, for example, A uses 3 lines.) Use the number of straight lines needed to make each of these letters to partition C into three sets.

2. Let $D = \{1, 2, 3, 4, 5, 6, 7, 8, 9, 10, 11, 12, 13\}$. When an integer is divided by 4, there is a remainder of 0, 1, 2, or 3. Use the remainder to partition D into 4 sets.

3. Every integer can be written as a product of primes. (An integer is a prime if its only divisors are 1 and itself.) For example, $60 = 2 \cdot 2 \cdot 3 \cdot 5$ is a factorization of 60 into primes using 4 terms. Use the number of terms in such factorizations to partition $U = \{4, 5, 6, 7, 8, 9, 10, 11, 12, 13, 14, 15, 16\}$.

4. Let $U = \{10, 11, 12, 13, 14, 15, 16, 17, 18, 19, 20, 21, 22, 23, 24, 25\}$. When an integer is divided by 3, there is a remainder of 0, 1, or 2. Use the remainder to partition U into 3 sets.

5. Let U be a universal set containing 60 elements. Suppose C and D are subsets of U with $\mathbf{n}(C) = 28$, $\mathbf{n}(D) = 24$, and $\mathbf{n}(C \cap D) = 11$. Compute the followng:

   (a) $\mathbf{n}(C \cup D)$,

   (b) $\mathbf{n}(C' \cap D)$,

   (c) $\mathbf{n}(C' \cap D')$,

   (d) $\mathbf{n}(C' \cup D')$.

6. Let A and B be subsets of U. Suppose $\mathbf{n}(U) = 50$, $\mathbf{n}(A) = 18$, $\mathbf{n}(B) = 23$, and $\mathbf{n}(A \cap B') = 12$. How many elements are in $A \cap B$? How many elements are in $(A' \cap B)'$?

7. For the sets A, B, and C in Sample Problem 2 of the text, use Figure 7 to determine the number of elements in the following sets:

(a) $A' \cup B' \cup C'$,

(b) $(A \cap B) \cup C$,

(c) $B \cap (A \cup C)$.

8. Using the data in Sample Problem 3 of the text, answer the following questions:

   (a) How many rooms contain exactly two of these three kinds of appliances?

   (b) How many rooms contain at most one of the three?

   (c) How many rooms contain at least one of the three?

9. Let A, B and C be subsets of the universal set U. Suppose $\mathbf{n}(U) = 100$, $\mathbf{n}(A) = 20$, $\mathbf{n}(B) = 40$, $\mathbf{n}(C) = 30$, $\mathbf{n}(A \cap B) = 10$, $\mathbf{n}(A \cap C) = 5$, $\mathbf{n}(B \cap C) = 15$, and $\mathbf{n}(A \cap B \cap C) = 3$. Determine the number of elements in the following sets:

   (a) $A \cap B' \cap C$,

   (b) $A' \cap B \cap C'$,

   (c) $A' \cap B' \cap C'$

10. Let D, E and F be subsets of the universal set U. Suppose $\mathbf{n}(U) = 50$, $\mathbf{n}(D) = \mathbf{n}(E) = 25$, $\mathbf{n}(F) = 31$, $\mathbf{n}(D \cap E) = 13$, $\mathbf{n}(D \cap F) = 17$, $\mathbf{n}(E \cap F) = 15$, and $\mathbf{n}(D \cap E \cap F) = 6$. Determine the following:

    (a) $\mathbf{n}(D \cap E' \cap F')$

    (b) $\mathbf{n}(D' \cap E \cap F)$

    (c) $\mathbf{n}(D \cup E \cup F)$.

11. In a class of 150 seventh graders, 60 have had chicken pox, 50 have had measles, and 40 have had mumps. Moreover, 30 have had both chicken pox and measles, 15 have had both chicken pox and mumps, 10 have had both measles and mumps, 5 have had all three.

    (a) How many have not had any of these diseases?

(b) How many have had at most one?

(c) How many have had at least two?

12. A baseball card collector has 200 different baseball cards. He has 95 outfielders, 50 New York Yankees, and 45 members of the Hall of Fame. Moreover, he has 15 Yankees who are members of the Hall of Fame, 35 Yankee outfielders, 25 outfielders who are members of the Hall of Fame, and 5 Yankee outfielders who are members of the Hall of Fame.

    (a) How many cards does he have which are neither Yankees, outfielders, nor members of the Hall of Fame?

    (b) How many Hall of Fame members does he have who are neither Yankees nor outfielders?

    (c) How many outfielders does he have who are either members of the Hall of Fame or Yankees but not both?

13. At a regional warehouse there are 95 fifteen cubic foot refrigerators in stock. There are 72 self-defrosting refrigerators, 44 have built-in ice makers, and 22 have a built-in cold water dispenser. There are 30 self-defrosting refrigerators with an ice-maker, and 12 self-defrosting refrigerators with a cold water dispenser. Finally, there are 16 refrigerators with both an ice maker and a cold water dispenser, but 6 of these 16 are not self-defrosting.

    (a) How many refrigerators are self-defrosting but have neither an ice maker nor a cold water dispenser?

    (b) How many refrigerators have at most two of these three additional features?

    (c) How many refrigerators have none of these three additional features?

14. Let A, B, and C be subsets of U. Suppose $\mathbf{n}(A \cap B' \cap C) = 1$, $\mathbf{n}(A \cap B' \cap C') = 2$, $\mathbf{n}(A \cap B \cap C') = 3$, $\mathbf{n}(A) = 10$, $\mathbf{n}(A' \cap B' \cap C) = 5$, $\mathbf{n}(C) = 16$, $\mathbf{n}(A' \cap B) = 13$ and $\mathbf{n}(B' \cap C') = 26$. Draw an appropriate Venn diagram, and, in each of the regions, enter the number of elements in that region.

15. Let A and B be subsets of U with $\mathbf{n}(U) = 62$, $\mathbf{n}(A) = 24$, $\mathbf{n}(B) = 30$, and $\mathbf{n}(A' \cap B') = 22$. Find $\mathbf{n}(A' \cap B)$ and $\mathbf{n}(A' \cup B')$.

16. More than two inches of rainfall was measured at the airport in each of 8 months last year, and in 7 months last year the measured rainfall was less than six inches. How many months last year did the rainfall measure between two and six inches?

17. A school has 120 students of whom 45 take Math, 44 take English, 25 take Math and English, 21 take English and History, 23 take Math and History, 10 take Math and English and History, and 36 take none of the three subjects.

    (a) How many students take History?

    (b) How many students take Math but not History?

    (c) How many students take exactly one of these three subjects?

18. In the sixth grade of a school there are 39 students. A survey is taken regarding condiments for hot dogs. The results are: 22 like mustard, 17 like ketchup, 13 like pickle relish, 4 like both ketchup and pickle relish, 7 like mustard and pickle relish but not ketchup, 4 like mustard and ketchup but not pickle relish, and 8 like only mustard. Fill in the Venn diagram for this situation.

    (a) How many do not like any of these three condiments on their hot dog?

    (b) How many like ketchup and pickle relish but not mustard?

    (c) How many like exactly one of these 3 on their hot dog?

    (d) How many like exactly two of these 3 on their hot dog?

19. A survey of 100 people is made concerning which newspaper they read. They are asked whether they read the Times, the Post or the Bugle. The surveyors find that 10 people read none of these papers, no one is found who reads both the Times and the Bugle, 30 people read the Times, 30 people read the Bugle, 23 people read both the Post and the Bugle and 10 people read both the Times and the Post. How many people read only the Bugle? How many people read the Post?

20. Given that G and H are subsets of T and $\mathbf{n}(T) = 80, \mathbf{n}(H') = 38, \mathbf{n}(G' \cap H') = 8, \mathbf{n}(G) = 36$, find $\mathbf{n}(H)$.

21. You have 56 friends each one of whom enjoys at least one of the following sports: swimming, water skiing and skin diving. Specifically, 36 enjoy water skiing, 36 enjoy skin diving, 19 enjoy swimming and skin diving, 20 enjoy swimming and water skiing, 21 enjoy water skiing and skin diving, and 11 enjoy all three sports. How many enjoy only swimming?

22. You are given the following data about the subsets A, B, and C of U: $\mathbf{n}(B \cap C \cap A') = 4$, $\mathbf{n}(B \cap C) = 9$, $\mathbf{n}(A \cap C) = 5$, $\mathbf{n}(C) = 12$, $\mathbf{n}(A \cap B) = 6$, $\mathbf{n}[(A \cup B \cup C)'] = 8$, $\mathbf{n}[(B \cup C)'] = 10$, $\mathbf{n}(U) = 30$. Draw an appropriate Venn diagram and in each of the eight regions write the number of elements in that region.

23. A pizza shop sold 68 pizzas yesterday, of which 30 had mushrooms, 39 had onions, 36 had anchovies, 15 had mushrooms and onions, 20 had onions and anchovies, 7 had mushrooms and onions and anchovies, and 12 had mushrooms and anchovies but no onions.

    (a) How many had mushrooms and anchovies?

    (b) How many had none of the three toppings?

    (c) How many had at least two of the toppings?

1.3 Sample Spaces

The setting for the mathematical study of probability is a sample space. By definition a *sample space* is the set of all possible outcomes of a random phenomenon. The random phenomenon can be an activity, game, experiment, etc. whose outcome is unpredictable. Among the simplest and most familiar random phenomena are tossing a coin, rolling a die, and spinning a wheel of chance (roulette wheel or something similar). In none of these can we predict the outcome ahead of time. The subsets of a sample space are called *events* because they represent something that might happen. An event consisting of one and only one element from the sample space is called a *simple event*, and two events whose intersection is the empty set are said to be *disjoint or mutually exclusive*. Elements of a sample space will often be referred to as points or sample points.

Once a sample space has been constructed for a random phenomenon, the next step is to assign probabilities to the events. The probability of an event is to be a measure of the likelihood of the event actually happening. When this is complete, the random phenomenon will have been replaced by a mathematical model consisting of a sample space (set) and probabilities (numbers) assigned to the events (subsets). In this section we will construct a variety of sample spaces before introducing the fundamental rules of probability in the next section.

EXAMPLE 1. The natural sample space for the roll of a standard six-sided die is $S = \{1, 2, 3, 4, 5, 6\}$. Here the numbers in S represent the different possible number of dots on the top face when the die comes to rest.

EXAMPLE 2. A slightly more complicated situation is tossing a coin three times. For each toss there are two possible outcomes - heads or tails. Let H stand for heads and T for tails. The outcome of heads on the first toss and tails on the second and third is naturally represented by HTT. Systematically writing down all possible outcomes in this way produces the sample space

$$S = \{HHH, HHT, HTH, THH, HTT, THT, TTH, TTT\}.$$

It is now easy to replace a verbal description of an event with a precise list of the elements in it. For example, if E is to be the event of obtaining exactly one head in the three tosses of the coin, then

$$E = \{HTT, THT, TTH\}.$$

Given a random phenomenon, it may be possible to construct more than one sample space depending upon what is considered to be an outcome. Constructing a suitable sample space must be done with an eye on the problem(s) to be solved. Usually the more information a sample space contains the more useful it is, which the next example illustrates.

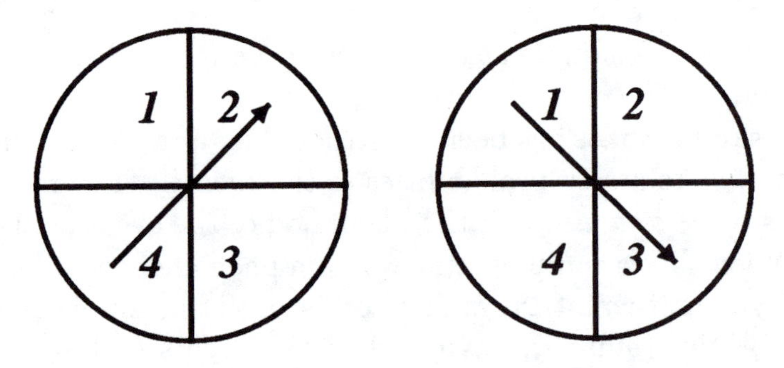

Figure 1

Sample Problem 1. *Two wheels of chance with 4 numbers on each one are spun. One wheel is green and the other is red, and the numbers on both of them are 1, 2, 3, and 4. (See Figure 1.). Construct a suitable sample space and list the elements in each of the following events:*

(a) The sum of the two numbers the two wheels stop at is even.

(b) The green wheel stops at 1.

SOLUTION: If we view the outcome of the random phenomenon to be the number at which the green wheel stops plus the number at which the red wheel stops, then the smallest outcome is 2, the largest is 8, and we have the sample space

$$S_1 = \{2, 3, 4, 5, 6, 7, 8\}.$$

Moreover, the answer to (a) is clearly

$$\{2, 4, 6, 8\}.$$

When we turn to part (b), a serious difficulty arises. The outcome could be 4 with the green wheel at 1 and the red at 3, or with both wheels at 2. Should 4 be listed in the answer to (b) or not? This is a hopelessly ambiguous situation; the sample space S_1 is simply not suitable for answering (b). To answer (b) we must construct a sample space which keeps track of both wheels individually.

Let (x, y) denote an ordered pair of numbers, that is, x is the first number and y is the second. In particular, $(2, 7) \neq (7, 2)$ because the order is different, Using the numbers at which the green wheel stops for the first number and the number at which the red wheel stops for the second number, the set

$$S_2 = \left\{ \begin{array}{l} (1,1),(1,2),(1,3),(1,4), \\ (2,1),(2,2),(2,3),(2,4), \\ (3,1),(3,2),(3,3),(3,4), \\ (4,1),(4,2),(4,3),(4,4) \end{array} \right\}$$

is a sample space for spinning both wheels. The event "the green wheel stops at one" can now be written $\{(1,1),(1,2),(1,3),(1,4)\}$. Moreover, we can also use S_2 to answer part (a). The event "the sum of the outcomes is even" corresponds to the subset

$$\{(1,1),(1,3),(2,2),(2,4),(3,1),(3,3),(4,2),(4,4)\}$$

of S_2. In summary, the sample space S_2 with its 16 elements contains a lot more information about the result of spinning the two wheels than the smaller sample space S_1. □

For a simple experiment like rolling a die or spinning a roulette wheel constructing a sample space is easy. However, many experiments proceed in stages with the possible outcomes at the second stage depending on the outcome at the first stage. For example, in some board games using dice, players are given an extra roll of the dice when they roll doubles. A tree diagram is a useful device for determining all the possible outcomes in such a situation.

The idea of a *tree diagram* is to draw lines from the results of each stage to the possible results at the next stage. We begin with a dot (vertex) and draw lines emanating from it to dots representing all the possible results of the first stage. Now repeat the process for each first stage result making the tree branch. Of course, this construction can be repeated until nothing further can happen and the total number of branches will depend on what happened at all the previous stages. Then all possible outcomes of the experiment appear as the paths from the root (initial dot) to the tips of the twigs (final dots). In particular, the number of these paths or branches is equal to the number of final dots. A family tree is a familiar example of a tree diagram. Further examples will best explain the details for constructing and using tree diagrams.

Sample Problem 2. *What are the possible scenarios in a best of three sets tennis match?*

SOLUTION: When the first player wins a set, write down the letter A and let B denote that the second player won the set. Thus AA means that the first player won the first and second sets and the match is over. However, if the result of the first two sets is AB, then the match is not over and there are two possible outcomes for the third set. We will use a tree diagram to systematically write down all the possibilities. We start with two branches going from the initial dot to dots marked A and B. From each of them there are also two branches to dots labeled A and B because either player can win the second set as shown in Figure 2. (They would look more like trees drawn vertically, but they are easier to draw and label horizontally.) Since the match is over if the first player wins the first two sets, the branch from the first dot labeled A to the second dot labeled A can not branch further and terminates. In contrast the branch from A to B must again split into two branches going to dots labeled A and B where they terminate. Figure 2 shows the complete tree diagram and from it we read off the set of all possible scenarios or sample space by listing the labels on each path from left to right. Specifically

$$S = \{AA, ABA, ABB, BAA, BAB, BB\}$$

and the solution is complete. □

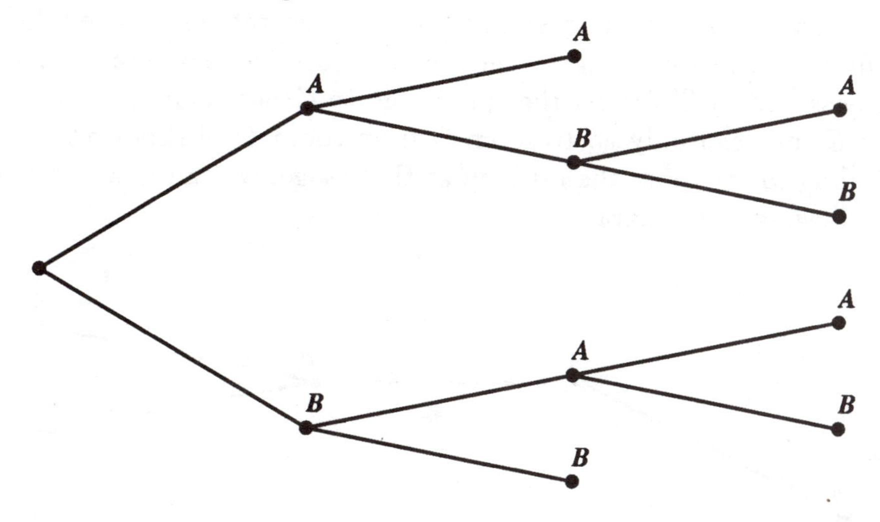

Figure 2

Sample Problem 3. *A rat is placed in the room labeled B of the maze shown in Figure 3 and allowed to roam. The experimenter keeps track of the next three rooms the rat enters. Construct a sample space for this experiment. How many different paths can the rat take visiting three rooms? What rooms can the rat be in at the end of the experiment? (In a more sophisticated experiment there would be food in one room and maybe an electric shock in another and the rat would be allowed to roam until it got the food or the shock.)*

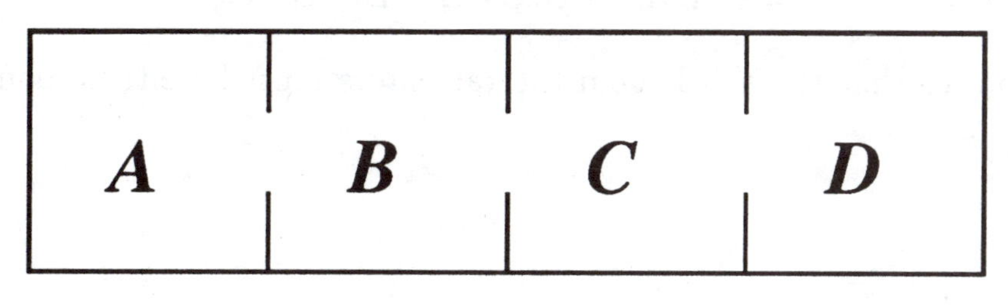

Figure 3

SOLUTION: From room B the rat can go next to either rooms A or

C, but once it enters room A it has no choice but to return to B. (Note that in the wording of the problem the rat was not explicitly prohibited from returning to the room it just left, so we must allow this possibility.) Similarly the rat can go to either room B or D from room C, but can only go to room C from room D. Labeling the dots according to the room the rat is in at that stage, we can now construct the following tree diagram:

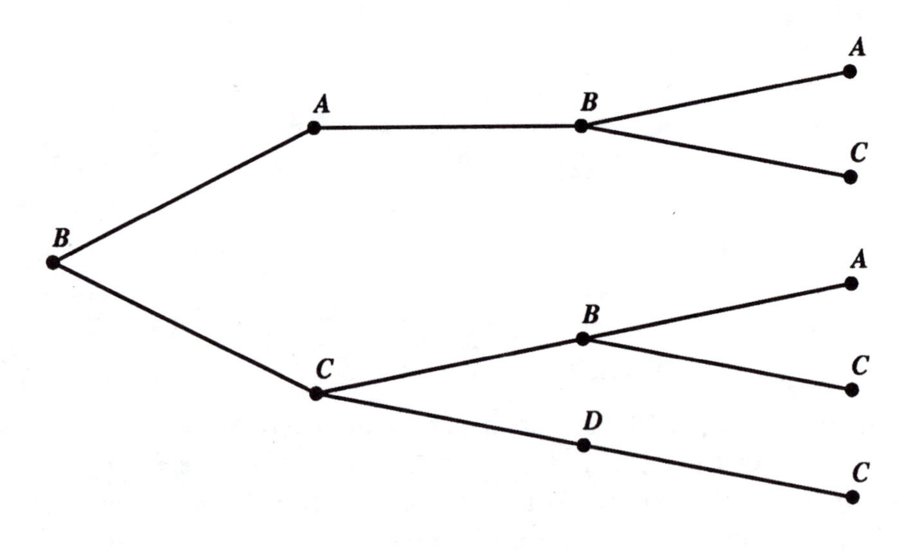

Figure 4

From the tree diagram we read off the sample space

$$S = \{ABA, ABC, CBA, CBC, CDC\}$$

and notice that the third room the rat enters must be either room A or C. □

Exercises

1. An experiment consists of both tossing a fair coin and rolling a fair die. Construct a sample space for this experiment. List the elements in the following events:

    (a) heads and an odd number on the die,

    (b) tails or an even number on the die,

    (c) neither heads nor a six on the die.

2. A particular brand of frozen pizza comes with any (including none and all) of the following extras: mushrooms, onions, pepperoni, and sausage. Instead of reading the labels you select a pizza at random. List the elements in a sample space describing this situation. List the elements in the following events:

    (a) You select a pizza with at least 3 extras,

    (b) you select a pizza with mushrooms,

    (c) you select a pizza with exactly 2 extras on it.

3. Suppose you have 2 extra tickets to a football game; one is for a seat on the 50 yd. line and the other behind the goal posts. Four friends would each like to buy one of them from you. To be fair, you decide who gets to buy which ticket by chance. Construct a sample space for this situation using A, B, C, and D as your friends' names. List the elements in the event either A or D gets the ticket on the 50 yd. line.

4. A box contains a one dollar bill, a two dollar bill, a five dollar bill, a ten dollar bill, and a twenty dollar bill. Construct a sample space for randomly drawing 2 bills from the 5 bills in the box. List the elements in the following events:

    (a) The total value of the bills selected is more than 15 dollars,

    (b) The total value of the bills selected is an even number of dollars.

5. A jury judging an art contest has four members and must select a chair and a recorder. They decide to make these selections by drawing names. Construct a sample space for this selection process.

6. A paint crew of 5 must be divided into two teams of 3 and 2 workers to paint a large and small room. Construct a sample space for a random selection of the two teams.

7. On a TV game show the amount of the first prize is determined by randomly selecting 4 numbered balls one after the other from a box containing them. The balls are numbered 1, 2, 4, and 8. The number on the first ball selected determines the first digit for the amount of the prize. For example, $8,421, the largest amount possible for the first prize, comes from drawing the balls in the order 8, 4, 2, 1. Construct a sample space for determining the amount of the first prize.

8. Use tree diagrams to determine all possible scenarios for a best of five sets tennis match at the U. S. Open and for the World Series (best of seven games).

9. A rat is placed in the middle room of a five room maze like the one in Figure 3. Construct a sample space for keeping track of the first four rooms the rat visits. Do you see a pattern here?

1.4 The Fundamentals of Probability

If probabilities are to be useful they must arise naturally in many situations. There are three broad categories of probabilities that are very common and are classified by the way they are determined - subjective judgement, relative frequency, and counting. The following are illustrations of these three means of determining probabilities:

1. An avid basketball fan who has closely followed the regular season in the Atlantic Coast Conference (ACC) might believe that the probability of Maryland winning the ACC tournament is 3/10, the probability of North Carolina winning is 2/10, etc.. These probabilities would be based entirely on the fan's subjective judgment; another fan might see it very differently.

2. A bottling machine at a beverage company occasionally fails to properly cap a bottle. What is the probability that a soda bottled by this machine will not be properly capped? Suppose someone observes the output of this machine for one hour. If in this hour the machine fills 5000 bottles of soda and 50 of them are not properly capped, then the relative frequency of bottles not properly capped is $50/5000 = 1/100 = 0.01$. It would be reasonable to assume that this is typical of the machine's operation and that the probability that this machine will not properly cap a bottle is 0.01. Thus the relative frequency provides an experimental or empirical means of assigning probabilities to events.

3. A group of friends attend a banquet together and there is a drawing for a door prize worth one hundred dollars. What is the probability that one of this group of friends will win the door prize? First count the number of people at the banquet; suppose there are 160 people there. Then count the number of people in this group; suppose there are 8. Then the probability that the winner is from this group of friends is 8 chances out of 160 or $8/160 = 1/20 = 0.05$. More generally counting the number of favorable outcomes and dividing by the the total number of possibilities is the classical notion of probability.

The main purpose of this section is to introduce mathematical probability in a unified and consistent way that includes the probabilities determined in these three natural ways - subjective judgement, relative frequency, counting. Since it would be futile to construct a theory for a multitude of special cases, a broad approach is needed. To develop such a unified theory of probability we must set down some basic rules and require all assignments of probabilities to obey these rules. They should be few in number, easy to state, and sensible. The theory will then be built up from these rules or axioms; consequently, all general facts about probability will apply to any specific assignment to events of probabilities which obey these rules.

Let S be a sample space and denote the probability of an event E in S by P(E). Probabilities of events in a sample space will always be required to obey the following rules or axioms:

PROBABILITY AXIOMS

Axiom 1. The probability of an event is always a non-negative number, that is

$$P(E) \geq 0$$

for all events E in S.

Axiom 2. The probability of the sample space is one, that is,

$$P(S) = 1.$$

Axiom 3. If $A_1, \ldots, A_n$ is a partition of the event E, then the probability of E is the sum of the probabilities of the events $A_1, \ldots, A_n$; that is,

$$P(E) = P(A_1) + P(A_2) + \ldots + P(A_n)$$

where $A_1, A_2, \ldots, A_n$ is a partition of E.

These rules certainly meet the criteria of being few in number and easy to state, but are they sensible? What follows is one brief argument for the reasonableness of each of them from among the many possible. First because most measurements (length, area, volume, weight) are not allowed to be negative, the measurement of the likelihood of an event taking place should also be greater than or equal to zero (Axiom 1). Since S consists of all possible outcomes, it is 100% certain that something in S will happen and changing 100% to a decimal gives S a probability of 1 (Axiom 2). If the weight of two different objects is known, then the total weight of the objects is the sum of their individual weights. Similarly if a certain event happens 10% of the time and another event which is incompatible with the first happens 15% of the time, then one or the other happens 10% + 15% = 25% of the time (Axiom 3).

If E and F are mutually exclusive events, then clearly E and F form a partition of $E \cup F$ and by Axiom 3 $P(E \cup F) = P(E) + P(F)$. Thus we have

$$P(E \cup F) = P(E) + P(F) \text{ when } E \cap F = \emptyset$$

which is often used instead of Axiom 3 as one of the basic axioms or rules of probability.

Axiom 3 is the probabilistic version of the Partition Principle which was the central idea in the previous section. Furthermore, Axiom 3 will be used to derive other useful probability formulas in the same way that the Partition Principle was used to derive counting formulas. The analogy between counting and probability is no accident; remember that calculating the probability of an event can be thought of as counting the frequency with which the event occurs.

There are three elementary consequences of the **Probability Axioms** that are especially important. They will be used in various ways through out this book and be referred to as Property 1, 2, or 3. To derive these basic properties let S be a sample space and suppose probabilities which obey Axioms 1, 2, and 3 have been assigned to the events in S. Consider any event E contained in S. Because E, E' is a partition of S

$$1 = P(S) = P(E) + P(E')$$

by Axioms 2 and 3. Subtracting $P(E)$ from both sides of this equation produces **Property 1**

$$\boxed{P(E') = 1 - P(E).}$$

Now let F be another event in S. Because $E \cap F$, $E \cap F'$ is a partition of E (see Figure 20), by Axiom 3

$$P(E) = P(E \cap F) + P(E \cap F').$$

Now subtracting $P(E \cap F')$ from our new equation gives **Property 2**

$$\boxed{P(E \cap F') = P(E) - P(E \cap F).}$$

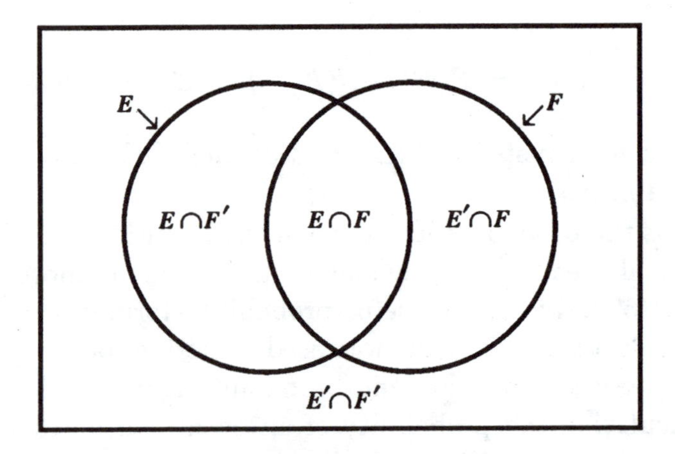

Figure 1

From Figure 1 it is also clear that $E \cap F'$, $E \cap F$, $E' \cap F$ is a partition of $E \cup F$, so again by Axiom 3

$$P(E \cup F) = P(E \cap F') + P(E \cap F) + P(E' \cap F).$$

Property 2 allows us to replace $P(E \cap F')$ by $P(E) - P(E \cap F)$ in this equation, and by analogy $P(E' \cap F)$ can be replaced by $P(F) - P(E \cap F)$.

The result of doing this is

$$P(E \cup F) = P(E) - P(E \cap F) + P(E \cap F) + P(F) - P(E \cap F).$$

Because

$$-P(E \cap F) + P(E \cap F) - P(E \cap F) = P(E \cap F)$$

we are left with simply $P(E) + P(F) - P(E \cap F)$ on the right hand side. Thus we have **Property 3**

$$\boxed{P(E \cup F) = P(E) + P(F) - P(E \cap F).}$$

Property 3 is analogous to the counting formula for unions,

$$\mathbf{n}(A \cup B) = \mathbf{n}(A) + \mathbf{n}(B) - \mathbf{n}(A \cap B),$$

and has the following similar interpretation: if you try to compute the probability of $E \cup F$ by adding the probabilities of E and F, then the probability of $E \cap F$ is measured twice and hence must be subtracted one time to obtain the probability of $E \cup F$. These formulas, especially Properties 1 and 3 will be used frequently.

Sample Problem 1. *Let S be a sample space for which the probabilities of events contained in S satisfy Axioms 1, 2 and 3. Let A and B be events contained in S with $P(A) = 0.4$, $P(B) = 0.45$, and $P(A \cap B) = 0.1$. Use the preceding formulas to compute $P(A')$, $P(A \cap B)$, $P(A \cup B)$, and $P(A' \cap B')$.*

SOLUTION: The first three calculations are:

$$P(A') = 1 - P(A) = 1 - 0.4 = 0.6$$

$$P(A \cap B') = P(A) - P(A \cap B) = 0.4 - 0.1 = 0.3$$

$$P(A \cup B) = P(A) + P(B) - P(A \cap B) = 0.4 + 0.45 - 0.1 = 0.75$$

For the last part recall that by de Morgan's formula $A' \cap B' = (A \cup B)'$ and hence

$$P(A' \cap B') = P[(A \cup B)'] = 1 - P(A \cup B) = 1 - 0.75 = 0.25.$$

An alternative way to calculate these probabilities is to make a Venn diagram, write the probability of each cell in that cell of the diagram, and then add up the probabilities of the appropriate cells to answer each question. In this example we first write 0.1 in $A \cap B$. Since $P(A) = 0.4$, we must write 0.3 in $A \cap B'$ to make the probabilities in the A circle add up to 0.4. Similarly, we must write 0.35 in $A' \cap B$. Finally, to get the sum of all the probabilities to equal 1 we must write 0.25 in $A' \cap B'$. Now to compute $P(A')$ we add up all the probabilities outside the A circle; that is, using Axiom 3, $P(A') = 0.35 + 0.25 = 0.6$ which agrees with the earlier answer.

Because $A \cap B'$ and $A' \cap B'$ are cells we can read their probabilities off the diagram as $P(A \cap B') = 0.3$ and $P(A' \cap B') = 0.25$. Finally, $P(A \cup B) = 0.3 + 0.1 + 0.35 = 0.75$. □

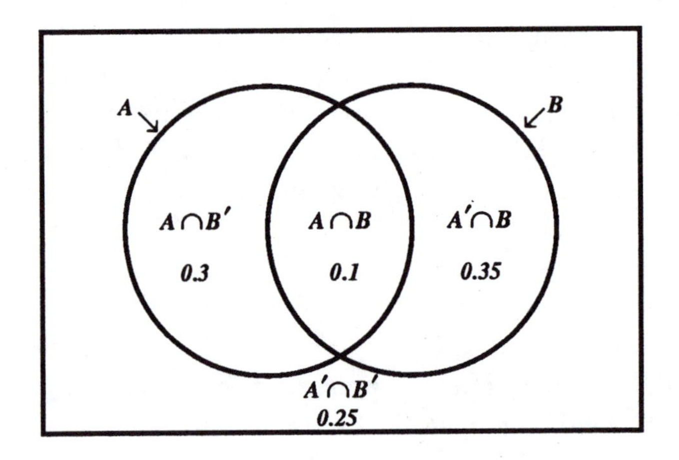

Figure 2

In the next example we will need to know $P(\emptyset)$. We would certainly expect it to be 0 and it is 0 because $S' = \emptyset$ and hence $P(\emptyset) = P(S') = 1 - P(S) = 1 - 1 = 0$ by Axiom 2.

Sample Problem 2. *Let D, E, and F be events in a sample space S, with $P(D) = 0.5$, $P(E) = 0.3$, $P(F) = 0.4$, $P(D \cap E) = 0.15$, $P(D \cap F) = 0.2$, and $P(E \cap F) = 0.1$. Moreover, suppose $D' \cap E \cap F = \emptyset$.*

Compute the following probabilities: $P(D \cap E \cap F')$, $P(D \cap E' \cap F')$, *and* $P(D \cap E \cap F)$. *What is the probability that at most one of these events occurs?*

SOLUTION: This problem will be solved by making a Venn diagram for three sets, writing each cell's probability in the appropriate place on the diagram, then reading off the answers. The reasoning will be very similar to what was used in Sample Problems 2 and 3 in the previous section.

The key to getting started is that $P(D' \cap E \cap F) = 0$ because $D' \cap E \cap F = \emptyset$. Since $D \cap E \cap F$ and $D' \cap E \cap F$ form a partition of $E \cap F$, $P(E \cap F) = 0.1$ implies $P(D \cap E \cap F) = 0.1$ because $P(D' \cap E \cap F) = 0$. Similarly $0.15 = P(D \cap E) = P(D \cap E \cap F) + P(D \cap E \cap F')$ implies $P(D \cap E \cap F') = 0.05$ and $0.2 = P(D \cap F) = P(D \cap E \cap F) + P(D \cap E' \cap F)$ implies $P(D \cap E' \cap F) = 0.1$. The work thus far is shown in Figure 3.

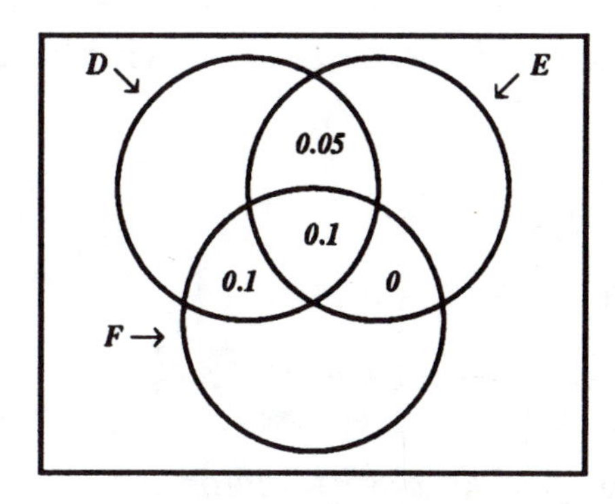

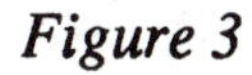
Figure 3

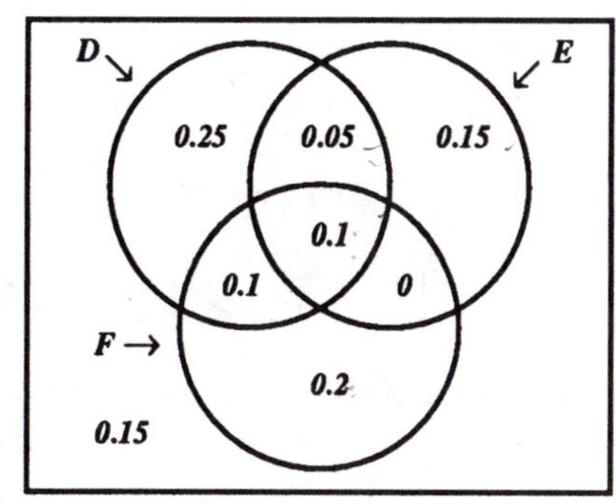

Figure 4

Next use the partition of D consisting of $D \cap E \cap F$, $D \cap E' \cap F$, $D \cap E \cap F'$, $D \cap E' \cap F'$ to see that

$$0.5 = P(D) = 0.1 + 0.1 + 0.05 + P(D \cap E' \cap F')$$

which implies

$$P(D \cap E' \cap F') = 0.25.$$

Applying the same idea to E and F produces

$$P(D' \cap E \cap F') = 0.15$$

$$P(D' \cap E' \cap F) = 0.2.$$

To force the sum of the probabilities of all eight cells to equal 1, we must have

$$P(D' \cap E' \cap F') = 1 - (0.1 + 0.05 + 0.1 + 0 + 0.25 + 0.2 + 0.15) = 0.15$$

which completes Figure 4. From Figure 4 we see that

$$P(D \cap E \cap F') = 0.05$$

$$P(D \cap E' \cap F') = 0.25$$

$$P(D \cup E \cup F) = 0.1 + 0.05 + 0.1 + 0 + 0.25 + 0.2 + 0.15 = 0.85.$$

To answer the last question note that at most one means none ,$D' \cap B' \cap C'$, or exactly one, which consists of the three cells lying inside precisely one of the circles. Hence

$$P(D' \cap E' \cap F') + P(D \cap E' \cap F') + P(D' \cap E \cap F') + P(D' \cap E' \cap F)$$

$$= 0.15 + 0.25 + 0.2 + 0.15 = 0.75.$$

is the probability of at most one of these event s happening □

Although it has been implicit in the Sample Problems that probabilities are never greater than one, it is worth explicitly showing that this property is another consequence of the Axioms. As in our earlier discussion let E and F be events in a sample space S and assume the probabilities of events in S obey Axioms 1, 2, and 3. Suppose $F \subset E$. It follows that $E \cap F = F$ and $E \cap F$ can be replaced by F in the Property 2 formula

$$P(E \cap F') = P(E) - P(E \cap F)$$

giving us

$$P(E \cap F') = P(E) - P(F).$$

Because $0 \leq P(E \cap F')$ by Axiom 1 it follows that $0 \leq P(E) - P(F)$ or $P(F) \leq P(E)$ by adding $P(F)$ to both sides. Therefore,

$$P(F) \leq P(E) \text{ when } F \subset E.$$

In other words, if an event is enlarged, its probability will not decrease. Since for any event E, $\emptyset \subset E \subset S$, it follows that

$$0 \leq P(E) \leq 1.$$

This last inequality is useful in a negative way; any time the calculation of a probability produces a negative number or a number greater than one, there is a mistake in the data or in the calculation.

Throughout the rest of these notes, any time probabilities are mentioned it will always be understood that they obey Axioms 1, 2, and 3. Consequently Properties 1, 2, and 3 hold along with the inequalities just derived and $P(\emptyset) = 0$. In summary, the following facts will always be at our disposal for any events E and F in a sample space S:

$P(E) \geq 0$
$P(S) = 1$
$P(E) = P(A_1) + P(A_2) + \ldots + P(A_n)$ when $A_1, A_2, \ldots, A_n$ is a partition of E
$P(E') = 1 - P(E)$
$P(E \cap F') = P(E) - P(E \cap F)$
$P(E \cup F) = P(E) + P(F) - P(E \cap F)$
$P(\emptyset) = 0$
$P(F) \leq P(E)$ when $F \subset E$
$P(E) \leq 1.$

Exercises

1. A student types a 5,000 word term paper and subsequently finds 30 words incorrectly typed. What is the probability that the student mistypes a random word when typing a paper?

2. If you flip a fair coin 4 times, what is the probability of getting heads exactly 3 times? What is the probability of getting heads at least three times?

3. While golfing with two friends, you mentally assign probabilities of 0.5 that friend A will have the lowest score and 0.4 that friend B will have the lowest score. What probability must you then assign to winning the game yourself?

4. A spinner for a board game is numbered from 1 to 5 in equal segments. What is the probability that the spinner stops at an even number? At a number bigger than 1?

5. If A_1, A_2, A_3, A_4 is a partition of C with $P(A_1) = 1/15$, $P(A_2) = 2/15$, $P(A_3) = 3/15$, $P(A_4) = 4/15$, what are the probabilities of the events C, C' and $A_2 \cup A_4$?

6. A student with important English and mathematics exams on the same day estimates that the probability he will pass the English exam is 0.7, the probability that he will pass the mathematics exam is 0.6, and the probability he will pass both exams is 0.2. What is wrong with this assignment of probabilities?

7. Let S be a sample space and let A and B be events in S with $P(A) = 4/5$, $P(B) = 2/5$, and $P(A \cap B) = 7/20$. Determine the following:

   (a) $P(A \cup B)$,

   (b) $P(A' \cap B')$,

   (c) $P(A' \cup B')$,

   (d) $P(A' \cap B)$.

8. Let A and B be events in a sample space S with $P(A) = 0.5$, $P(B) = 0.6$, and $P(A \cap B') = 0.3$. Compute the following:

   (a) $P(A \cap B)$,

   (b) $P(B')$,

   (c) $P(A \cup B)$,

   (d) $P(A \cup B')$.

9. Let A and B be events in a sample space. Which of the following statements about the probabilities of A and B are impossible?

   (a) $P(A) = 0.1$, $P(B) = 0.2$, and $P(A' \cup B') = 0.7$

   (b) $P(A) = 1/12$, $P(B) = 5/12$, $P(A \cap B) = 0$, and $P(A' \cap B) = 5/12$

   (c) $P(A) = 0.4$, $P(B) = 0.3$, and $P(A' \cap B') = 0.2$.

10. Someone tells you the probability that the metro bus is on time is 0.63, the probability it is late is 0.16, and the probability it is early is 0.23. Is this possible? If not, why is it impossible?

11. For the sets D, E and F in Sample Problem 2 of the text use Figure 4 to calculate the probability of the events $D \cap (E \cup F)$, $F \cup (D \cap E)$, $D' \cup E' \cup F'$.

12. Let E and F be events in a sample space S with $P(E) = 0.4$, $P(F) = 0.7$ and $P(E' \cap F') = 0.05$. Compute the following probabilities:

    (a) $P(E \cup F)$,

    (b) $P(E \cap F)$,

    (c) $P(E' \cap F)$,

    (d) $P(E' \cup F)$.

13. Let A, B, and C be events in a sample space with $P(A) = 7/20$, $P(B) = 6/20$, $P(C) = 8/20$, $P(A \cap B) = 1/20$, $P(A \cap C) = 2/20$. Moreover, B and C are mutually exclusive. Compute the probabilities of the following events:

(a) $A \cap B' \cap C$,

(b) $A' \cap B \cap C'$,

(c) $A \cup B \cup C$,

(d) $(A \cup B) \cap C'$.

14. If a pay phone is selected at random along a certain turnpike, the probability that it is out of order is $1/4$, the probability that there is no phone book is $5/8$, and the probability it is out of order and there is no phone book is $3/16$. What is the probability that there is a phone book and the phone is out of order? What is the probability of selecting a working phone with a phone book?

15. A person is selected at random from a large sample of Americans. The person selected has a college education with probability 0.3, is over 35 years old with probability 0.5, and is married with probability 0.35. Further, the person selected has a college education and is married with probability 0.12, has a college education and is over 35 with probability 0.16, and is both over 35 and married with probability 0.22. Finally, the probability that the person selected does not have a college education but is married and over 35 is 0.18.

    (a) What is the probability that an unmarried person under 35 without a college education is selected?

    (b) What is the probability of selecting a married person without a college education?

    (c) What is the probability that the person selected is unmarried, over 35, and has a college education?

16. In a certain state a political party is trying to win the governorship, a U.S. Senate seat, and the majority of seats in the state legislature. Using the work of a pollster, they estimate their chances as follows: the probability of winning the governorship is 0.35, the probability of winning the Senate seat is 0.35, the probability of winning a majority in the state legislature is 0.5, the probability of winning both the governorship and the Senate seat is 0.16,

the probability of winning a majority in the state legislature and the governorship is 0.15, the probability of winning a majority in the state legislature and the Senate seat is 0.25, and finally the probability of winning only the governorship is 0.14. What is the probability that they will succeed in at most one of these endeavors? What is the probability they will fail in at least one? What is the probability they will succeed in exactly two of the three?

17. Let A, B, C be events in a sample space S. Suppose that $P(A) = 0.53$, $P(B) = 0.34$, $P(C) = 0.28$, $P(A \cap B') = 0.33$, $P(B \cap C) = 0.13$, $P(A \cap C) = 0.18$ and $P(A' \cap B \cap C) = 0.06$. Compute the following probabilities:

    (a) $P(A \cap B)$,

    (b) $P(A' \cap B' \cap C')$,

    (c) $P[(A \cup B) \cap C']$.

18. An employee credit union makes car loans, personal loans and home improvement loans available to its members. In a random sample of its members the probability that a member had a car loan was 0.60, the probability that a member had a personal loan was 0.45, and the probability that a member had a home improvement loan was 0.22. The probability that a member had both car and personal loans was 0.25 and the probability of having all three types was 0.03. The probability of having both home improvement and personal loans was 0.05 while the probability of only having a home improvement loan was 0.10.

    (a) What is the probability that a member had exactly two types of loans?

    (b) What is the probability that a member had at most one type of loan?

19. The Venn diagram given below (Figure 5) for the events E, F, and G in the sample space S shows the probability of each of its cells. Use it to compute the probabilities of the following events:

(a) $E \cup F \cup G$,

(b) $E' \cup F' \cup G'$,

(c) $F \cup G$,

(d) $(E \cup F) \cap G'$.

20. The following is a direct quote from the Feb. 23, 1984 Diamondback, the student newspaper of the University of Maryland:

"According to the 1982 survey, the proportion of students relying:

on no grants, loans or earnings was 25 percent,
only on earnings was 24 percent,
on grants, loans and earnings was 11 percent,
on grants and earnings was 12 percent,
on loans and earnings was 9 percent,
only on grants was 8 percent,
only on loans was 5 percent,
on grants and loans was 5 percent."

Use this data to complete a Venn diagram. There is an interpretation problem you will have to resolve. There is also a defect in the data. Can you find it? Now answer the following questions:

(a) Ignoring this defect, calculate the probability that a student relies on grants (not necessarily exclusively on grants).

(b) What is the probability that a student has both a grant and a loan?

(c) What is the probability that a student relies on at least two of these sources for financial support?

21. Let A and B be events in a sample space S. Suppose that $P(A' \cap B) = 0.1$, $P(A \cup B) = 0.6$, and $P(B) = 0.4$. Compute the following probabilities:

(a) $P(A' \cap B')$,

(b) $P(A)$.

22. Rex was interrupted in the middle of a Venn diagram problem. His unfinished diagram is shown below (Figure 6). He still has the following bits of information to use: $P(R') = 0.78; P(T) = 0.37; P(Q) = 0.26$. Complete the Venn diagram by filling in the appropriate probabilities.

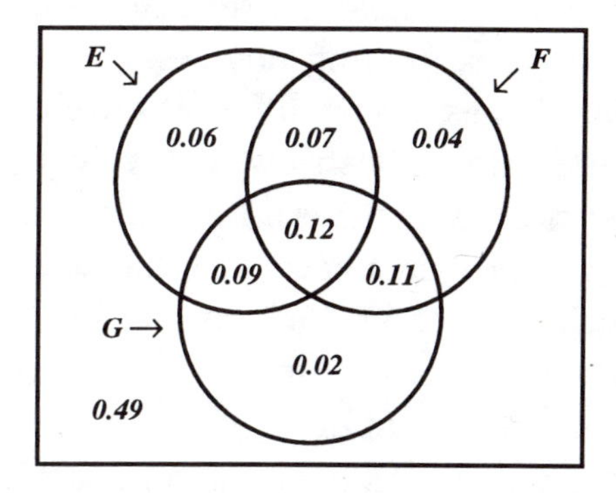

Figure 5

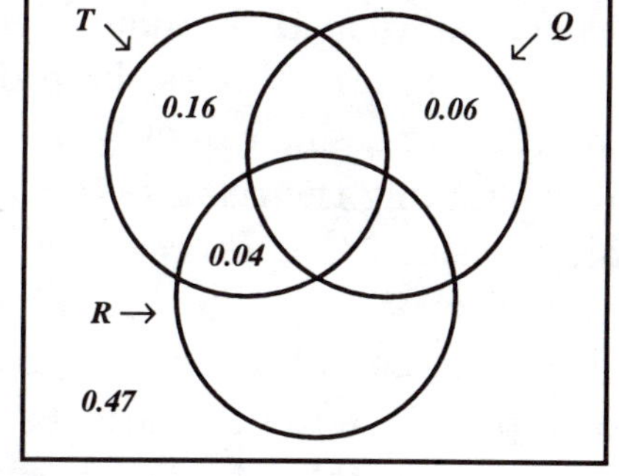

Figure 6

1.5 The Classical Concept of Probability

Man's fascination with games of chance and gambling is as old as civilization itself. The familiar six sided cubic die originated 4,000 years ago, and for more than a 1,000 before that the astragalus was used in games of chance. The astragalus is the ankle bone of a sheep or goat which lands in one of four positions when thrown. Polished marked versions have been found in Egyptian and near eastern archeological sites (circa 3,600 B. C.) and identified as gaming devices. Because of the irregularities of bones the probabilities of the different outcomes of a specific astragalus can only be determined by throwing it many times and calculating the relative frequencies. In contrast, for a die it is natural to assume that its regularity implies that each face has the same probability, namely 1/6. Probabilities like this which can be calculated by counting and dividing will be the focus of this section; they are also the origin of the mathematical study of probability.

It was not until the seventeenth century that any serious attempts were made to replace lady luck with reason. The principle characters involved in the creation of probability were Chevalier de Mere, a gentleman gambler; Blaise Pascal, a mathematician and theologian; and Pierre Fermat, a jurist and mathematician. In 1654 de Mere challenged his friend Pascal to explain why he won more by betting even money on rolling at least one six in 4 rolls of a die than on rolling double sixes at least once in 24 rolls of a pair of dice? In a series of letters Pascal and Fermat unraveled this problem and many similar ones and began the subject of probability.

The axiomatic approach which we discussed in the previous section is a comparatively recent development having been discovered by the Russian mathematician, A. N. Kolmogorov around 1933. Although probability had thrived in the almost three hundred years between de Mere's challenge and Kolmogorov's axioms, it lacked the simple elegant unity that the axioms gave it. The axioms were used to introduce probability in the previous section to provide a coherent approach to probabilities of all sorts, but now the historical gap between the original seventeenth century idea of probability and the modern axiomatic approach must be bridged.

The original notion of probability, which is still very useful, was the ratio of the number of favorable possibilities to the total number of possibilities. We want to fit this definition with its gambling overtones into the framework of our general study of probability. The total number of possibilities is, of course, the number of elements in the sample space. What are the favorable possibilities? If you are playing a game of chance, certain possible outcomes favor you and the rest favor your opponent. So favorable means different things to different players, but the favorable possibilities will always be some subset of all possible outcomes. In other words, the favorable possibilities will always be an event. Thus this definition of probability amounts to the ratio of the number of elements in an event E to the number of elements in the sample space S and depends only on counting the elements in different sets. This definition of probability can be written

$$\boxed{P(E) = \frac{\mathrm{n}(E)}{\mathrm{n}(S)}}$$

and will be referred to as *the classical formula for probability.* Note that this is nonsense unless the sample space S is finite.

The next step is to see if the formula

$$P(E) = \frac{\mathrm{n}(E)}{\mathrm{n}(S)}$$

really deserves to be called a formula for probability. Recall our earlier dictum that we will only consider probabilities that obey Axioms or rules 1, 2, and 3. So we must verify that these fundamental rules will be obeyed when this formula is used to assign probabilities to events. Since $\mathrm{n}(E) \geq 0$ and $\mathrm{n}(S) > 0$, clearly $P(E) = \mathrm{n}(E)/\mathrm{n}(S) \geq 0$ and Axiom 1 holds. Next $P(S) = \mathrm{n}(S)/\mathrm{n}(S) = 1$, which is Axiom 2. Suppose $A_1, A_2, \ldots, A_n$ is a partition of the event E. From the Partition Principle we know that

$$\mathrm{n}(E) = \mathrm{n}(A_1) + \mathrm{n}(A_2) + \ldots + \mathrm{n}(A_n).$$

If we divide both sides of this equation by $\mathrm{n}(S)$, we get

$$\frac{\mathrm{n}(E)}{\mathrm{n}(S)} = \frac{\mathrm{n}(A_1)}{\mathrm{n}(S)} + \frac{\mathrm{n}(A_2)}{\mathrm{n}(S)} + \ldots + \frac{\mathrm{n}(A_n)}{\mathrm{n}(S)}$$

Therefore,

$$P(E) = P(A_1) + P(A_2) + \ldots + P(A_n)$$

and Axiom 3 is valid. Consequently the classical formula is an acceptable way to assign probabilities to events in a finite sample space and all the general facts about probability can be used as needed. We turn now to examining its usefulness as a method for measuring the likelihood of an event.

The classical formula has one very definite limitation. If we use the classical formula and E is a simple event, that is, $E = \{a\}$ where a is one element of S, then $\mathbf{n}(E) = 1$ and

$$P(E) = \frac{1}{\mathbf{n}(S)}$$

which is independent of the choice of the element a from S. Thus simple events all have the same probability or are equally likely. Consequently, the classical formula cannot be used unless there is reason to believe that simple events are equally likely.

Conversely, if we know that the simple events are equally likely in a finite sample space, then the classical formula is forced upon us. To see this, suppose that the common probability of every simple event is p. All the simple events partition S into $\mathbf{n}(S)$ pieces. Hence

$$1 = P(S) = \underbrace{p + p + \ldots + p}_{\mathbf{n}(S) \text{ times}} = \mathbf{n}(S) \cdot p$$

and dividing by $\mathbf{n}(S)$ gives

$$p = \frac{1}{\mathbf{n}(S)}.$$

By the same token the simple events contained in the event E partition E into $\mathbf{n}(E)$ pieces with probability $p = 1/\mathbf{n}(S)$. As above

$$P(E) = \underbrace{p + p + \ldots + p}_{\mathbf{n}(E) \text{ times}} = \mathbf{n}(E) \cdot p = \mathbf{n}(E) \cdot \frac{1}{\mathbf{n}(S)} = \frac{\mathbf{n}(E)}{\mathbf{n}(S)}$$

which is the classical formula.

Consider a normal six-sided die and let $S = \{1, 2, 3, 4, 5, 6\}$. Because of the symmetry of the cube, there is no reason to believe one side is favored over another. So the probability of each simple event should be 1/6 and the classical formula is the natural way to assign probabilities. Additional support for this view can be obtained by doing a frequency study. To do this the die is rolled many times, say 6,000, and each outcome is recorded. Unless the die has been tampered with in some way each number will have occurred about, but not necessarily exactly, 1,000 times. For example, the results might be

S	Occurrences	Relative Frequency
1	995	0.16583
2	1002	0.16700
3	1008	0.16800
4	991	0.16516
5	997	0.16616
6	1007	0.16783

Since the decimal equivalent of 1/6 is 0.16666..., this would be strong evidence that each simple event should be assigned the probability of 1/6. In many situations the frequency approach to probability can be used in this way to corroborate the use of the classical formula.

Now suppose someone added 2 dots to the face with 1 dot, and produced a die having two sides with 3 dots and one each with 2, 4, 5, and 6 dots. If we use the sample space $S_1 = \{2, 3, 4, 5, 6\}$, we would expect the simple event 3 to have probability $2/6 = 1/3$ and the other simple events to have probability 1/6 each. Hence the classical formula could not be used. If the die was altered in a less obvious way, a good frequency study would reveal the bias and could be used to assign probabilities to simple events.

Although the classical formula cannot be used with careless abandon, there are many situations where it is the right way to assign probabilities to events. This includes almost all games of chance provided no one has tampered with the equipment. The key word that is often used to indicate that simple events are equally likely is "fair" as in a fair coin or a fair die. The classical formula also applies when names are drawn from a hat, cards from a deck, or balls from an urn. The key phrases are then "random selection" or "selected at random".

There is one other pitfall that must be avoided; the correct use of the classical formula is dependent upon the construction of a suitable sample space. It is easy to get answers that are dead wrong by applying the classical formula to a sample which does not contain enough information.

Sample Problem 1. *A pair of fair dice is rolled. Compute the probability of each of the following events:*

(a) The sum of the numbers on the dice is 10. (the number on a die after it is rolled always refers to the number of dots on the top face).

(b) The sum of the numbers on the dice is 7 or 11.

(c) The sum is at most 5.

(d) The number on at least one of the dice is a 3.

SOLUTION: First a suitable sample space must be constructed. This will be done using ordered pairs of numbers in the same way that the second sample space for the two wheels of chance in Sample Problem 1 in Section 3 was constructed. For convenience assume one die is red and the other green. Then let the first number be the outcome for the red die and the second number the outcome for the green die. So $(5,2)$ means a 5 on the red die and a 2 on the green die. Systematically listing all possibilities produces the sample space shown below. Note that there are 36 elements in S. (This sample space will reappear many times in the remaining chapters.) The probability of each event will be computed by listing the elements in the event, counting, and dividing by 36. It will be convenient here and in similar problems to denote the events in parts (a), (b), (c), and (d) by the letters A, B, C, and D.

$$S = \left\{ \begin{array}{l} (1,1),(1,2),(1,3),(1,4),(1,5),(1,6), \\ (2,1),(2,2),(2,3),(2,4),(2,5),(2,6), \\ (3,1),(3,2),(3,3),(3,4),(3,5),(3,6), \\ (4,1),(4,2),(4,3),(4,4),(4,5),(4,6), \\ (5,1),(5,2),(5,3),(5,4),(5,5),(5,6), \\ (6,1),(6,2),(6,3),(6,4),(6,5),(6,6) \end{array} \right\}$$

(a) First list the elements in S with sum equal to 10 as

$$A = \{(4,6),(5,5),(6,4)\}$$

and then use the classical formula

$$P(A) = \frac{\mathrm{n}(A)}{\mathrm{n}(S)} = \frac{3}{36} = \frac{1}{12}.$$

(b) The probability of getting 7 or 11 is calculated in the same way, except now the list of elements in B consists of those pairs whose sum is either 7 or 11,

$$B = \{(1,6),(2,5),(3,4),(4,3),(5,2),(6,1),(5,6),(6,5)\}$$

$$P(B) = \frac{\mathrm{n}(B)}{\mathrm{n}(S)} = \frac{8}{36} = \frac{2}{9}$$

(c) Observe that requiring the sum to be at most 5 is the same as requiring the sum to be 2, 3, 4, or 5 and then proceed as in the previous part to get

$$C = \left\{ \begin{array}{l} (1,1),(1,2),(2,1),(1,3),(2,2), \\ (3,1),(1,4),(2,3),(3,2),(4,1) \end{array} \right\}$$

$$P(C) = \frac{\mathrm{n}(C)}{\mathrm{n}(S)} = \frac{10}{36} = \frac{5}{18}.$$

One of the benefits of listing the elements in S as a 6 by 6 array is that the sum is constant on ascending diagonals. The elements in the events A, B, and C were all determined by listing the elements on the appropriate diagonals.

(d) The event in the last part is described in terms of the individual dice not their sum. The elements in D consist of the elements of S in either the third column or the third row, so

$$D = \left\{ \begin{array}{l} (1,3),(2,3),(3,3),(4,3),(5,3),(6,3), \\ (3,1),(3,2),(3,4),(3,5),(3,6) \end{array} \right\}$$

and

$$P(D) = \frac{\mathrm{n}(D)}{\mathrm{n}(S)} = \frac{11}{36}.$$

Solution is now finished □

Odds provide another approach to probability. The *odds* against an event E is the ratio of the number of unfavorable outcomes to the number of favorable outcomes which is usually written $a : b$. Since the number of unfavorable outcomes is precisely $\mathrm{n}(E')$, the odds against E is the ratio $\mathrm{n}(E') : \mathrm{n}(E)$. Because $\mathrm{n}(S) = \mathrm{n}(E) + \mathrm{n}(E')$ we can recover the probability of E from $a : b$, the odds against E as follows:

$$P(E) = \frac{\mathrm{n}(E)}{\mathrm{n}(S)} = \frac{\mathrm{n}(E)}{\mathrm{n}(E) + \mathrm{n}(E')} = \frac{b}{a+b}.$$

EXAMPLE 1. If the odds against E are $7 : 3$, then

$$P(E) = \frac{3}{7+3} = 0.3$$

Sample Problem 2. *In the game of craps the shooter wins on the first roll of a pair of dice if he or she rolls seven or eleven. What are the odds against the shooter winning on the first roll?*

SOLUTION: Let E be the event the shooter wins on the first roll. Using the sample space in Sample Problem 1,

$$E = \{(1,6),(2,5),(3,4),(4,3),(5,2),(6,1),(5,6),(6,5)\}$$

and $\mathbf{n}(E) = 8$. It follows that $\mathbf{n}(E') = 28$ and the odds against E are $28 : 8$. Because 4 divides both 28 and 8, this ratio is the same as $7 : 2$. Both answers are correct, but ratios are usually reduced the same way fractions are put in lowest terms. □

Odds are usually used to express the payoff on a wager. (We will discuss the difference between house odds and true odds in Chapter 4.) If the odds against the bettor are $a : b$ which is the conventional way of giving odds to a bettor, then the bettor wins a dollars on a bet of b dollars.

EXAMPLE 2. Betting on an event E at $5 : 2$ odds means that you would win \$5.00 on a \$2.00 bet when E occurs and lose your \$2.00 when it does not occur.

Although dice provide a good familiar setting to illustrate the basic idea of the classical formula for probability, the use of this formula has a wide range of applications. Anytime an element of a finite set is randomly selected the classical formula applies. To show how random selection is linked to our earlier discussion of counting, we will return to Sample Problem 2 in Section 2. In this problem we were given data about the number of rooms in a 500 room dormitory that contained refrigerators, stereos, and televisions. Specifically,there are televisions in 175 rooms, stereos in 225 rooms, and refrigerators in 190 rooms. Furthermore, 90 rooms contain both a television and a stereo, 75 contain both a television and a refrigerator, and 85 contain both a stereo and refrigerator. Finally, 40 rooms contain a television and a stereo but no refrigerator. By carefully analyzing the data, we constructed the following Venn diagram (Figure 8 in Section 2):

We can now ask probabilistic questions about the same situation and use the classical formula to answer them.

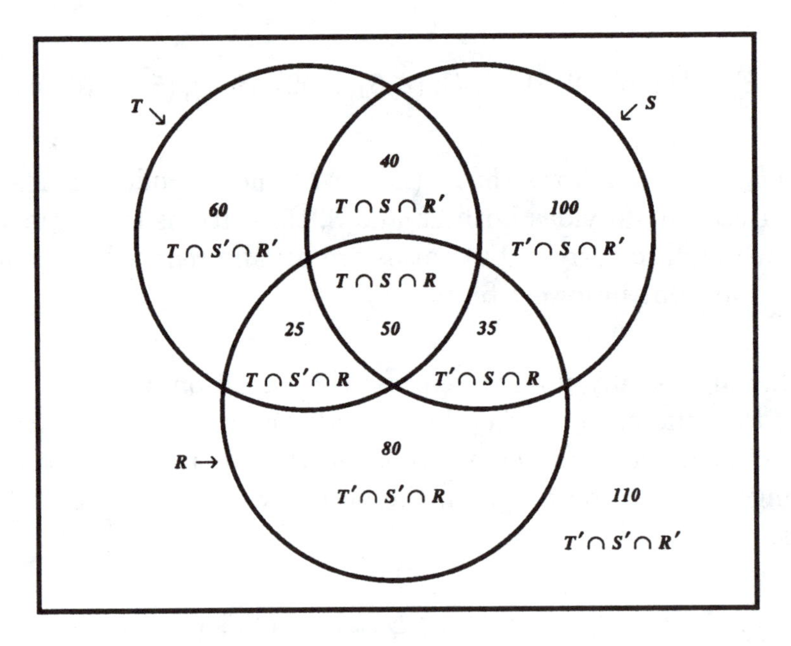

Figure 1

Sample Problem 3. *Suppose someone in the student life office randomly selected a room in this dormitory and inspected it. What is the probability that he or she will find a television or a stereo in the room? What is the probability of finding at most two of the three appliances - television, stereo, or refrigerator - in the room selected for inspection?*

SOLUTION: The sample space is just the set of all rooms in the dormitory and because the selection is random we can use the classical formula and the information in Figure ??. The first question asks for $P(T \cup S)$. Using the classical formula we get

$$P(T \cup S) = \frac{\mathrm{n}(T \cup S)}{\mathrm{n}(U)} = \frac{310}{500} = \frac{31}{50}.$$

For the second question let E be the event he finds at most two of the three kinds of appliances. Observe that E is the complement of finding all three or

$$E = (T \cap S \cap R)'.$$

Since

$$P(T \cap S \cap R) = \frac{\mathrm{n}(T \cap S \cap R)}{\mathrm{n}(U)} = \frac{50}{500} = 0.1$$

it follows that

$$P(E) = 1 - 0.1 = 0.9.$$

This is a standard trick. □

Sample Problem 4. *An urn contains 4 balls numbered 1, 2, 3, and 4. Two balls are randomly selected at once. (Someone closes their eyes, reaches into the urn, and pulls out two balls at the same time.) What is the probability that exactly one of the two balls selected will have an odd number on it?*

SOLUTION: First we construct a sample space by systematically listing all the possible outcomes:

$$S = \{1\&2, 1\&3, 1\&4, 2\&3, 2\&4, 3\&4\}.$$

Next we list the elements in the event "exactly one of the two balls selected has an odd number on it":

$$E = \{1\&2, 1\&4, 2\&3, 3\&4\}.$$

We can use the classical formula for probability to obtain

$$P(E) = \frac{\mathrm{n}(E)}{\mathrm{n}(S)} = \frac{4}{6} = \frac{2}{3}$$

and complete the solution. □

If the urn in Sample Problem 4 contained 25 balls numbered from 1 to 25, then the probability that exactly one of the two balls selected has an odd number on it could in principle be computed in the same way. However, to make a list of elements in S would be an exceedingly dull and tedious task. What we need are some techniques for counting the objects in S and E without making lists. The entire next chapter will be devoted to more sophisticated counting techniques and their applications to probability.

Exercises

1. Two wheels of chance are spun. The first is marked with the numbers 1, 2, 3, 4, and the second with 1, 2, 3. On both wheels each number is equally likely. List the elements in a sample space and compute the probabilities of the following events:

   (a) Exactly one of the numbers is odd.

   (b) Both numbers are even.

   (c) The sum of the numbers is 5.

   (d) The sum of the numbers is at most 5.

2. Suppose you have a fair die and four cards numbered 1, 2, 3, 4. You roll the die and draw a card. Construct a suitable sample space to describe the outcomes of this random phenomenon. Use the classical formula to compute the probability of the following events:

   (a) The number on the card is bigger than the number on the die.

   (b) The sum of the numbers is 6.

   (c) At least one of the numbers is 3.

   (d) The sum of the numbers is at most 4.

3. Among 100 new cars delivered to a dealer, 75 have power windows, 85 have dual air bags, 80 have ABS brakes, 65 have power windows and dual air bags, 60 have power windows and ABS brakes, 70 have dual air bags and ABS brakes, and 55 have all three. Compute the probability that a car selected at random from this lot has the following equipment:

   (a) At least one of the three options.

   (b) Only dual air bags.

   (c) Only dual air bags and power windows.

   (d) Exactly two of the three options.

4. A question on a History exam gives you 4 Danish names and asks you if any (including all or none) were tenth century Danish kings. Being caught totally unprepared for this question, you guess and hope. List the elements in a sample space describing this situation. For convenience use A, B, C, and D to denote the 4 names you are given. Suppose the correct answer is precisely A and B. Assuming the simple events in the sample space are equally likely, compute the following:

   (a) The probability that you write down at least 2 incorrect names on your paper.

   (b) The probability that exactly 1 of the names you write down is correct.

5. Two fair four-sided dice (tetrahedrons) are rolled. The sides of each one are marked with one, two, three, or four dots respectively, and the outcome of rolling one of them is the number of dots on the side facing the table.

   (a) What is the probability of doubles (same outcome on both dice)?

   (b) What is the probability that the sum of the two outcomes is at most 5?

   (c) What is the probability of getting a 4 on at least one of the two dice?

6. An urn contains a penny, a nickel, a dime, a quarter, a half dollar, and a Susan B. Anthony dollar. Two of these coins are randomly selected at once without replacement. List the elements in a sample space and compute the following probabilities:

   (a) The total value of the two coins selected is at most 30 cents.

   (b) The total value of the two coins selected is an integral multiple of ten cents (in other words, the same value as some whole number of dimes).

   (c) The total value of the two coins selected is more than 15 cents.

(d) The more valuable of the two coins selected is worth exactly five times the value of the other coin selected.

7. Let E and F be subsets of U with $\mathbf{n}(U) = 50, \mathbf{n}(E) = 35, \mathbf{n}(F) = 25$, and $\mathbf{n}(E' \cap F') = 5$. If an object is selected at random from U, what is the probability that an object from $E \cap F$ will be selected?

8. A wheel of chance used at a fair is numbered from 1 to 40 inclusive and each number is divided into three equal parts. The center part for each number is painted red and the outer two parts are painted white. Thus it is possible to bet on red 25 or white 25 etc.. What is the probability that the wheel which is assumed to be fair will stop at the red part of an even number? What is the probability that the wheel will stop at the white part of a multiple of 5?

9. A fair 8-sided die (octahedron) whose sides are numbered from one to eight is rolled. What is the probability that the outcome is a multiple of 3? What is the probability that the outcome is a positive power of 2?

10. Three people each toss a fair coin. What is the probability that exactly two of the outcomes will be the same?

11. A number is chosen at random from the set $\{1, 2, 3, \ldots, 99, 100\}$. What is the probability that the number chosen is larger than 88 or a multiple of 5? (This is one question, not two.)

12. Three glasses containing different wines are randomly put in a line for tasting. Two of the wines are American and one is French. What is the probability that the American wines are tasted in succession?

13. Let A and B be subsets of U with $\mathbf{n}(U) = 200, \mathbf{n}(A \cap B') = 70, \mathbf{n}(A' \cap B) = 20$, and $\mathbf{n}(A) = 120$. An element is randomly selected from U. What is the probability that:

    (a) An element from B is selected?

(b) An element from $A \cup B$ is selected?

14. Two wheels of chance are spun. The first wheel has the numbers 1, 2, 3, 4 (each number is equally likely) and the second has the numbers 1, 2 (again, each number is equally likely).

    (a) Construct a sample space for this experiment.

    (b) Find the probability that the number on the first wheel equals the number on the second wheel.

    (c) Find the probability that the sum of the numbers on the two wheels is less than or equal to 5.

15. An element is selected at random from a set U containing 20 objects. Given subsets A and B of U with $P(A) = 0.25, P(B) = 0.3$, and $P(A \cap B) = 0.1$, determine $\mathrm{n}(A \cup B)$.

16. If the probability of winning a game is $0.4 = 4/10$, what are the odds against winning this game? What are the odds for winning the game?

17. What are the odds against rolling doubles with a pair of dice? What are the odds against rolling seven or eleven?

18. At a horse racing track one of the regulars tells you he thinks the odds against a certain horse winning are 8:5. What probability is he assigning to this horse winning the race? If someone else told you the odds against this horse winning are 9:11, now what probability is being assigned to this horse winning?

19. Calculate $P(E)$ when the odds against E are 3:4.

1.6 Review Problems

1. Let A and B be subsets of U with $\mathbf{n}(U) = 60$. Given that $\mathbf{n}(A) = 23$, $\mathbf{n}(B) = 31$, and $\mathbf{n}(A' \cap B') = 17$, calculate $\mathbf{n}(A \cap B)$.

2. Suppose E and F are events in a sample space S with $P(E) = 0.4$, $P(F) = 0.7$, and $P(E' \cap F) = 0.55$. What are the probabilities of the events $E \cup F$ and $E' \cup F'$.

3. In a small sample of 60 students attending a university, it was found that 9 were living off campus, 36 were undergraduates, and 3 were undergraduates living off campus.

   (a) Find the number of students in this sample that were undergraduates, were living off campus, or both.

   (b) Find the number of students that were undergraduates living on campus.

   (c) Find the number of students that were graduate students living on campus.

4. A survey of students taking English or mathematics indicates that 45 students are taking mathematics and 41 are taking English. How many students could there be altogether in this survey? If it is also known that 12 students are taking both English and mathematics, how many students would there be altogether?

5. There are 38 automobiles in a lot: 20 of these are foreign cars, 24 have a standard transmission and 25 have tape players. Of the 13 foreign cars with tape players 6 also have a standard transmission. Eleven foreign cars do not have a standard transmission, and 8 cars have a standard transmission but no tape player.

   (a) Use a Venn diagram to record these data.

   (b) How many cars have no tape player and are not foreign?

   (c) How many cars have a standard transmission or are foreign?

   (d) How many cars have exactly one of these features?

6. Let $E = \{2,3,4,5,6,7,8,9,10\}$ and use the number of divisors of the integers in E to partition E into 3 sets. For example, the divisors of 6 are $1,2,3$, and 6, so 6 has 4 divisors and would be put with other integers in E having 4 divisors.

7. Draw a Venn diagram containing three sets clearly marked A, B, C. Shade the set
$$(A \cap B) \cup (A' \cap B' \cap C).$$

8. An urn contains four balls: two black balls numbered 1 and 2, and two white balls also numbered 1 and 2. An experiment consists of selecting two of these balls at once randomly.
   (a) Describe a sample space for this experiment. (Hint: each outcome listed in your sample space shoud be as likely to occur as any other outcome. Use the notation $B1$ to denote the black ball numbered 1 etc., to keep track of the four balls.)
   (b) What is the probability that the two selected balls are of different colors?
   (c) What is the probability that the numbers on the two selected balls are the same?

9. A universal set of 32 elements contains two subsets A and B such that $A' \cap B' = \emptyset$. There are 20 elements in A and 25 elements in B. How many elements are in each of the following sets:
   (a) $A \cap B$
   (b) $A' \cap B$
   (c) $A' \cup B'$?

10. Let $U = \{r,s,t,u,v,w,x,y,z\}$, $A = \{s,t,v,x,z\}$, $B = \{s,u,v,y,z\}$, $C = \{t,u,v,w\}$, and $D = \{r,s,t,x,y,z\}$. List the elements in the following sets:
   (a) $[(A \cup B) \cap C] \cap D'$
   (b) $(A' \cap C) \cup (C \cap D)$.

11. Suppose A, B, and C are subsets of the universal set U with $\mathbf{n}(U) = 54$, $\mathbf{n}(A) = 7$, $\mathbf{n}(B) = 16$, $\mathbf{n}(C) = 20$, $\mathbf{n}(A \cap B) = 3$, $\mathbf{n}(A \cap C) = 2$, $\mathbf{n}(B \cap C) = 6$, and $\mathbf{n}(A' \cap B \cap C') = 8$. Determine the number of elements in each cell of a Venn diagram for these three sets. (Hint: To find the number of elements in the center cell use $\mathbf{n}(A \cap B) = 3$ and $\mathbf{n}(B \cap C) = 6$ simultaneously with $\mathbf{n}(B) = 16$.) How many elements are in $A \cap B \cap C$? How many elements are in $A' \cap B' \cap C'$?

12. Let G and H be events in a sample space S with $P(G) = 1/2$, $P(H) = 1/4$, and $P(G \cap H) = 1/6$. Compute the probability of the events
    (a) $G \cup H$,
    (b) $G' \cup H'$,
    (c) $(G \cup H) \cap (G' \cup H')$.

13. Construct a sample space for tossing a fair coin 4 times. Use the total number of heads on the 4 tosses to partition the sample space into five subsets. Assuming simple events are equally likely, compute the probability of each set in the partition.

14. Make a Venn diagram for three sets labeled P, Q, and R and shade $(P \cup Q \cup R) \cap (P' \cup Q' \cup R')$.

15. Let $U = \{1,2,3,4,5,6,7,8\}$, $E = \{1,2,3,4,5,6\}$, $F = \{4,5,6,7,8\}$, and $G = \{2,3,4,5,6,7\}$. Find $[(E \cap F) \cup G']'$ and $(F \cap G) \cap (G \cup E)'$.

16. A candy dish contains 100 pieces of fudge of which half contain nuts. There are 45 semi-sweet pieces, 21 large semi-sweet pieces, and 6 large semi-sweet pieces without nuts. Altogether there are 35 large pieces and the rest are small. Finally, there are 20 large pieces with nuts and 25 semi-sweet pieces with nuts. Suppose you select a piece at random.
    (a) What is the probability of selecting a small semi-sweet piece with nuts?
    (b) What is the probability of selecting a small semi-sweet piece?

(c) What is the probability of selecting a small piece without nuts?

17. An urn contains a penny, a nickel, a dime, a quarter, and a half dollar. Without replacement two coins are randomly selected one after the other.

    (a) List the elements in a sample space describing this experiment.

    (b) What is the probability that the first coin selected is worth less than the second one?

    (c) What is the probability that the second coin selected is worth five times as much as the first one?

18. Suppose A, B, and C are events such that B and C are mutually exclusive, $P(A) = P(B) = P(C) = 0.25$, $P(A \cap C) = 0.15$, and $P(A \cap B) = 0.1$. Determine

    (a) $P(A \cup B \cup C)$,

    (b) $P(A \cap B')$,

    (c) $P(B \cup (A \cap C))$.

19. Use Venn diagrams to decide whether or not

$$(A \cup B) \cap C' = A \cup (B \cap C').$$

20. A survey of 50 college students is taken regarding fast food restaurants. The results are : 9 like Roy Rogers but not Wendy's, 7 like Roy Rogers and McDonald's, 13 like McDonald's but not Wendy's, 8 like McDonald's and Wendy's, 18 like Wendy's, 5 like all three, and finally 1 likes Roy Rogers and Wendy's but not McDonald's. A student is selected at random from these 50 students. Compute the probability of the following events:

    (a) A student is selected who likes Roy Rogers and McDonald's but not Wendy's,

    (b) a student is selected who does not like any of McDonald's, Roy Rogers, or Wendy's

(c) a student is selected who likes at least two of the three.

21. A "Wheel of Fortune" is divided evenly into eights with each section numbered consecutively 1, 2, 3, 4, 5, 6, 7, 8. The wheel is spun twice and the number on which the wheel stops each time is noted.

    (a) How many possible outcomes are there?

    (b) What is the probability that the wheel stops on the number 3 on the first spin and 8 on the second spin?

    (c) What is the probability that the wheel stops on an even number on both spins?

    (d) What is the probability that the sum of the numbers on which the wheel stops is not greater than 4?

22. Urn 1 contains three balls: two black balls numbered 1 and 2, and a green ball numbered 3. Urn 2 contains three balls: two green balls numbered 1 and 3, and a black ball numbered 2. An experiment consists of drawing a ball from each urn. Construct a suitable sample space and determine:

    (a) the probability the two balls have the same colors,

    (b) the probability they have the same numbers,

    (c) the probability they have the same colors or numbers,

    (d) the probability the number on the first ball is not greater than the number on the second,

    (e) the probability balls of different color are drawn and that the number on the green ball is the larger,

    (f) the probability that at least one green ball is drawn.

23 Given the universal set $U = \{1, 2, 3, 4, 5, 6, 7, 8, 9\}$ and the subsets $A = \{1, 3, 5, 6, 8\}$ and $B = \{4, 5, 6, 7\}$, list the elements in each of the following: (a) $(A' \cup B') \cap (A \cap B')$ (b) $(A \cap B') \cup (A \cup B)'$.

24. A drawer contains 3 pairs of gloves - a green pair, a yellow pair, and a blue pair. Two gloves are randomly selected at the same

time. Construct a sample space for this situation. What is the probability that both gloves selected are left handed? What is the probability of selecting a left handed and a right handed glove but not of the same color?

25. Draw two Venn diagrams containing the sets E, F, and G clearly marked. In the first shade $E' \cup (F \cap G)$ and in the second shade $(E' \cup F) \cap (E' \cup G)$. Use your Venn diagrams to decide whether or not it is true that $E' \cup (F \cap G) = (E' \cup F) \cap (E' \cup G)$.

26. Thirty people go for lunch. Eighteen order salad, thirteen do not order soup, and three order neither soup nor salad. How many people order both soup and salad?

27. Let A, B, and C be subsets of U with $\mathbf{n}(U) = 200, \mathbf{n}(A) = 106, \mathbf{n}(B) = 84, \mathbf{n}(A \cap B) = 50, \mathbf{n}(A \cap C) = 26, \mathbf{n}(B \cap C) = 30, \mathbf{n}(A' \cap B' \cap C') = 28,$ and $\mathbf{n}(A \cap B \cap C') = 30$. An object is selected at random from U. What is the probability of each of the following events?

    (a) An element from $A \cap B \cap C$ is selected?

    (b) An element from C is selected?

    (c) An element from $A \cup (B \cap C)$ is selected?

28. Two wheels of chance are spun. The first is marked with the numbers 1, 2, 3 and the second with 3, 4, 5. On both wheels each number is equally likely. Find the probabilities of the following events:

    (a) Both numbers are the same.

    (b) There is at least one "3".

    (c) The sum of the two numbers is six.

    (d) The sum of the two numbers is no more than eight.

    (e) The sum of the two numbers is three.

29. Sixty actors gathered for a reunion. Twenty-six of these had acted in films, 30 in Shakespearian plays and 27 in musicals. Of the four

who had acted in musicals and films, all but one also acted in a Shakespearian play. Fifteen had appeared only in films and a total of 19 had played in exactly 2 of these kinds of productions.

(a) How many have not acted in any such productions?

(b) How many have acted in musicals or films?

(c) How many have acted in precisely one kind of production?

(d) What is the probability that a randomly selected actor would have played in all three kinds of productions? in just musicals?

30. Suppose you have 4 keys that appear identical but only one will open a certain lock. Construct two different sample spaces for randomly selecting and trying one key after another until the lock opens. (One sample space should be constructed with a tree diagram and contain considerably more information than the other one.)

31. Use a tree diagram to construct a sample space showing all possible outcomes of the following telephone survey: The first question asked is, "Are you a registered Democrat, Republican or neither?". Those responding Democrat are asked if they expect to vote for Clinton in 1996, and those answering yes are asked whether or not they will contribute financially to his campaign. For those responding Republican, the surveyor reads a list of three prominent Republican politicians and asks which one would be their first choice for President. No other questions are asked.

Chapter 2

Counting and Random Selection

Random selection and the classical formula for probability were discussed in the last section of the previous chapter, but their use was severely limited by a lack of techniques for solving all but the simplest counting problems. Recall that for random selection the probability of an event E is given by

$$P(E) = \frac{\mathrm{n}(E)}{\mathrm{n}(S)}$$

which is called the classical formula for probability. The use of this formula is limited only by our ability to count the number of objects in the event E and the sample space S; however, this can be very difficult in a complex situation.

In this chapter a number of basic counting methods will be developed and applied to probability problems involving random selection. By combining these new methods with ideas from Chapter 1, we will be able to make effective use of the classical formula and solve a wide range of rather sophisticated probability problems. The cornerstone of this program is the Multiplication Principle.

The Multiplication Principle, like the Partition Principle, is a self evident truth. Because of their self evident nature these two principles are useful mathematical ideas and lead to the solution of many counting problems. By recognizing and clearly stating these fundamental principles, our reasoning becomes more coherent.

2.1 The Multiplication Principle

Frequently we want to count things that are constructed by selecting one component from each of several different categories. For example, if you own 8 shirts and 5 pairs of pants, how many different shirt and pants combinations do you have to choose from when you dress in the morning? Of course the answer is $8 \cdot 5 = 40$, and the Multiplication Principle is a formal statement of this idea.

MULTIPLICATION PRINCIPLE

If an operation consists of m separate steps of which the first can be performed in k_1 different ways, the second in k_2 different ways, the third in k_3 different ways, etc., then the number of different ways the entire operation can be performed is given by:

$$k_1 \cdot k_2 \cdot k_3 \cdot \ldots \cdot k_m.$$

In simple situations a tree diagram provides a good pictorial view of the Multiplication Principle. If you picture any tree diagram in which the dots at each stage all split into the same number of branches, then the Multiplication Principle provides a means of calculating the number of final dots or what is the same thing, the number of paths from the initial dot on the left to a final dot on the right. Unfortunately as soon as the numbers involved get larger, making a tree diagram becomes unwieldy.

EXAMPLE 1. Flying from Washington, D. C. to Los Angeles with a stopover in Chicago but no change of airports, you can leave from any one of three airports (National, Dulles, and Baltimore-Washington International) stopover in Chicago at O'Hare or Midway, and fly into Los Angeles International or Ontario International in Los Angeles. How many choices are available? Figure 1 is a tree diagram in which the first stage is the departing airport, the second stage is the stopover, and the third stage is the arriving airport. Since all three first stage dots split into two and they in turn each split into two branches, by the Multiplication Principle there are $3 \cdot 2 \cdot 2 = 12$ branches going from the beginning to the end which is also plainly evident in Figure 1.

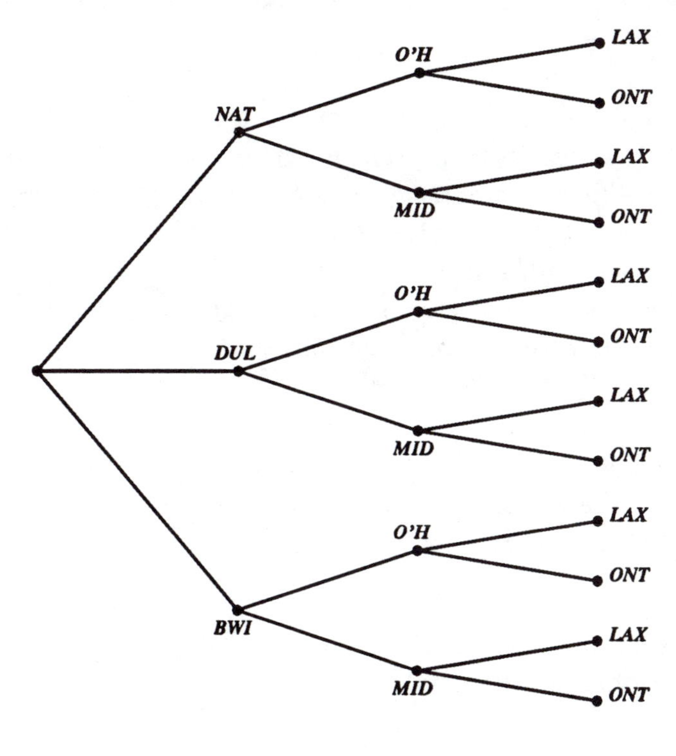

Figure 1

To show in greater detail how the Multiplication Principle works we examine an increasingly more complex situation. If you have 3 different kinds of bread and 5 different kinds of cheese, how many different kinds of cheese sandwiches can you make using only one kind of cheese and one kind of bread in each one? Here the operation is making the sandwich and there are 2 steps - selecting the bread and selecting the cheese. So $m = 2, k_1 = 3$, and $k_2 = 5$. By the Multiplication Principle the answer is $3 \cdot 5 = 15$.

If, in addition to the bread and cheese, you select one of 11 different kinds of meat, how many different kinds of sandwiches can you make? Now it is a 3 step operation ($m = 3$) with $k_1 = 3$, $k_2 = 5$, and $k_3 = 11$. Again by the Multiplication Principle the answer is $3 \cdot 5 \cdot 11 = 165$.

Finally, if we also have a choice of either mayonnaise or mustard, we have a 4 step operation with $k_1 = 3$, $k_2 = 5$, $k_3 = 11$, and $k_4 = 2$, and the number of different kinds of sandwiches we can now make is $3 \cdot 5 \cdot 11 \cdot 2 = 330$

Sample Problem 1. *Suppose a music lover wants to purchase components for a stereo system from a certain store. The store offers a choice of 10 different models of matched pairs of speakers, 5 different models of compact disk players, 8 different models of stereo receivers, 6 different models of head phones, and 3 different models of equalizers. Assuming the different models are all compatible with each other, how many different stereo systems consisting of one each of the above are there to choose from in the store?*

SOLUTION: The selection of each component is a step in the operation of purchasing the whole stereo system. So it is a 5 step operation with $k_1 = 10$, $k_2 = 5$, $k_3 = 8$, $k_4 = 6$, and $k_5 = 3$. The answer is

$$10 \cdot 5 \cdot 8 \cdot 6 \cdot 3 = 7,200.$$

by the Multiplication Principle. □

It may happen that each step in the operation can be performed in the same number of ways. In this case the formula $k_1 \cdot k_2 \cdot k_3 \cdot \ldots \cdot k_m$ from the Multiplication Principle can be written more simply as k^m where m is the number of steps in the operation and k is the common number of ways in which each step can be performed.

Sample Problem 2. *The call letters for radio stations usually begin with a W followed by three other letters. How many different call letters are possible if the first letter must be W and there are no restrictions on the three letters following the W?*

SOLUTION: Since the first letter must be a W, all allowed call letters can be obtained in a 3 step operation, namely, select the second, third, and fourth letters. Because there are 26 letters in the English alphabet, each of these three steps can be performed in 26 ways. So we can use

k^m as above and

$$26^3 = 17,576.$$

is the answer. □

The Multiplication Principle and the Partition Principle from Chapter 1 are the two fundamental principles of counting. All the counting problems in the previous chapter depended on the Partition Principle and all the new material on counting in this chapter will depend directly on the Multiplication Principle. Moreover, there are many problems whose solutions make use of both of these principles. The strategy in these problems is to find a partition $A_1, \ldots, A_n$ of the set C of objects being counted so that by the Partition Principle

$$\mathrm{n}(C) = \mathrm{n}(A_1) + \mathrm{n}(A_2) + \ldots + \mathrm{n}(A_n)$$

and so that each of the numbers $\mathrm{n}(A_1), \mathrm{n}(A_2), \ldots, \mathrm{n}(A_k)$ can easily be computed using the Multiplication Principle.

Sample Problem 3. *In the Morse code each signal for a letter consists of a series of one, two, three, or four dots and dashes in a definite order. For example, the signal for y is — · — —. How many such signals are possible?*

SOLUTION: Let B be the set of signals consisting of 1, 2, 3, or 4 symbols in a definite order where each symbol is a dot or a dash. Let A_1 denote the subset of such signals consisting of a single dot or dash; let A_2 denote the subset of signals consisting of 2 symbols; let A_3 denote those consisting of 3 symbols; and let A_4 denote those consisting of 4 symbols. This is a partition of B into four subsets and hence

$$\mathrm{n}(B) = \mathrm{n}(A_1) + \mathrm{n}(A_2) + \mathrm{n}(A_3) + \mathrm{n}(A_4).$$

Clearly $\mathrm{n}(A_1) = 2$. Constructing all the signals in A_2 is a two step operation and each step can be done in two ways - dot or dash. By the Multiplication Principle $\mathrm{n}(A_2) = 2^2 = 4$, and similarly $\mathrm{n}(A_3) = 2^3 = 8$ and $\mathrm{n}(A_4) = 2^4 = 16$. Combining these facts we have

$$\mathrm{n}(B) = 2 + 4 + 8 + 16 = 30.$$

So after you assign signals to the 26 letters you only have 4 extras. □

In a finite sample space S with equally likely simple events, the probability of an event E is given by

$$P(E) = \frac{\mathbf{n}(E)}{\mathbf{n}(S)}.$$

Thus the essential ingredient for computing probabilities in this context is the ability to count the elements in the sample space S and in the event E. The Multiplication Principle and other techniques in this chapter can often be used to determine $\mathbf{n}(S)$ and $\mathbf{n}(E)$. Although this approach does not require making a list of things in S which is usually tedious or impossible, it is still important to have a precise description of S and E in mind.

Sample Problem 4. *A quiz consists of five multiple choice questions and each question has four possible answers of which only one is correct. An unprepared student decides to guess. Assuming that with the student's method of guessing all possible answer sheets are equally likely, compute the probabilities of the following events:*

(a) The student answers every question correctly,

(b) The student answers them all incorrectly,

(c) The student answers at least one question correctly,

(d) The student answers only the first two questions correctly.

SOLUTION: Let S consist of all possible ways the entire quiz can be answered. Answering the quiz is a 5 step operation and each step can be performed in 4 ways. So by the Multiplication Principle we have

$$\mathbf{n}(S) = 4 \cdot 4 \cdot 4 \cdot 4 \cdot 4 = 4^5 = 1{,}024.$$

(a) This is the easiest part because there is only one correct solution to the entire quiz and the probability of obtaining it by guessing is

$$\frac{1}{1024} = 0.000977.$$

(b) Let E denote the event that every question is answered incorrectly. Since each question can be answered incorrectly in exactly 3 ways, we have

$$n(E) = 3^5 = 243$$

and

$$P(E) = \frac{n(E)}{n(S)} = \frac{243}{1,024} = 0.237.$$

(c) Frequently it is much easier to calculate the probability of the complement of an event than the probability of the event itself. The complement of L, the event that the student answers at least on question correctly, is the event that the student answers all of them incorrectly which is the event E from part (b). Therefore,

$$P(L) = 1 - P(L') = 1 - P(E) = 1 - 0.237 = 0.763.$$

(d) Let F be the event the student guesses correctly on the first 2 and incorrectly on the remaining 3. Again by the Multiplication Principle

$$n(F) = 1 \cdot 1 \cdot 3 \cdot 3 \cdot 3 = 27$$

and thus

$$P(F) = \frac{n(F)}{n(S)} = \frac{27}{1,024} = 0.0263.$$

Notice the difference in magnitude of the answers to (a) and (b). A perfect paper is very unlikely at roughly one in a thousand tries, but the chance of getting a zero on the quiz is close to one in four tries. □

In the last section of Chapter 1 we carefully listed all the elements in the sample space S for rolling a pair of dice. It contained the 36 pairs of integers (a, b) with a and b ranging from 1 to 6. It is now clear that the number 36 comes from the Multiplication Principle because rolling a pair of dice can be thought of as a two step operation - roll the red die recording the result and then do the same for the green die - and each step produces 1 of 6 possible results giving a total of $6 \cdot 6 = 36$ elements in S.

Rolling a pair of dice is obviously no different than rolling one die twice. So we would use the same sample space, except now the element $(5,3)$ of S would correspond to getting a 5 on the first roll and 3 on the second roll. Of course, the same observation applies to rolling three dice once or rolling one die three times, but now we need a bigger sample space. Instead of pairs like $(5,3)$ we need triples like $(5,3,4)$. Since there are six possibilities for each of the three numbers in the triple, by the Multiplication Principle there must be $6^3 = 216$ elements in the sample space for rolling a die three times. Similar reasoning applies to 4, 5, and more rolls of a die. With these ideas we can now solve de Mere's problem which stimulated the creation of probability by Pascal and Fermat. (See the beginning of Section 1.5.)

Sample Problem 5. *Is it better to bet even money on rolling a six at least once in 4 rolls of a die or on rolling double sixes at least once in 24 rolls of a pair of dice?*

SOLUTION: Betting even money means that the amount you can win is the same as the amount you can lose; if you bet one dollar you either win or lose exactly one dollar. Consequently all we need to do is calculate the probability of winning each bet and see which one is the larger.

First let S_1 be the sample space for rolling one die 4 times. We want to count the number of elements in S_1 without listing them. Reasoning as above S_1 consists of all quadruples of the integers from 1 to 6, like (1,1,1,1), (1,1,1,2), (1,1,1,3), etc. Since there are 6 possible outcomes for each roll, by the Multiplication Principle there must be $6^4 = 1,296$ elements in the sample space or $\mathbf{n}(S_1) = 6^4$.

Next let E be the winning event of at least one six in the 4 rolls. Computing $\mathbf{n}(E)$ is hard, but calculating $\mathbf{n}(E')$ is easy. Since E' is the event that no sixes occur in the four rolls, there are only 5 possible outcomes for each roll in E' and $\mathbf{n}(E') = 5^4$ again by the Multiplication Principle. Thus

$$P(E) \quad = \quad 1 - P(E') = 1 - \frac{\mathbf{n}(E')}{\mathbf{n}(S_1)}$$

$$\begin{aligned} &= 1 - \frac{5^4}{6^4} = 1 - \left(\frac{5}{6}\right)^4 \\ &= 1 - 0.4823 = 0.5177. \end{aligned}$$

To calculate the probability of winning the second bet a calculator is essential, but the ideas are the same as the one we just completed. Each time a pair of dice is rolled there are 36 possible outcomes. By analogy with a single die, when a pair of dice are rolled two, three or four times there are 36^2, 36^3 or 36^4 possible outcomes respectively. Rolling a pair of dice 24 times produces an enormous sample space S_2 containing 36^{24} elements.

Let F denote the winning event of double sixes at least once in the 24 rolls for the second bet. Again it is much easier to calculate $P(F')$ and use $P(F) = 1 - P(F')$. Excluding double sixes leaves 35 possible outcomes in each roll of the pair of dice. Following the same reasoning used to calculate the probability of winning the first bet, $\mathrm{n}(F') = 35^{24}$ and

$$\begin{aligned} P(F) &= 1 - P(F') = 1 - \frac{\mathrm{n}(F')}{\mathrm{n}(S_2)} \\ &= 1 - \frac{35^{24}}{36^{24}} = 1 - \left(\frac{35}{36}\right)^{24} \\ &= 1 - 0.5086 = 0.4914. \end{aligned}$$

The calculation $(35/36)^{24} = 0.5086$ was done with the "y^x" key which is on most calculators.

Notice that the probability of winning the first bet, $P(E) = 0.5177$, is slightly greater then the probability of winning the second bet, $P(F) = 0.4914$, and we conclude that the first bet is the better of the two. Although this small difference between the two bets would be unnoticeable in a few bets, it becomes significant when the bet is repeated hundreds of times. In fact, the first bet would be a winning bet over the long haul because it is slightly bigger than 0.5, and the second bet is a long term loser. □

The final example of this section uses its three main ideas, namely, the Multiplication Principle, the strategy of partitioning, and the calculation of probabilities by the classical formula.

Sample Problem 6. *Suppose a telephone number is randomly selected from one exchange, say 314. What is the probability that there is exactly one 0 in the last four digits, for example 314 - 7306.*

SOLUTION: Let S be the set of all possible phone numbers in one exchange. Since every such number begins with the same first three digits, we need only concern ourselves with the last 4 digits. Each of these last four digits can be 0, 1, 2, 3, 4, 5, 6, 7, 8, or 9, so there are 10 possibilities for each of them. Hence $\mathbf{n}(S) = 10^4 = 10,000$. Let E be the event such a phone number contains exactly one 0. Because the 0 can occur in any one of four places we partition accordingly. Let A_1 be the subset of S in which the first three digits are followed immediately be a 0 and then 3 non-zero digits. Define A_2, A_3, and A_4 similarly to complete the partitioning. There are 9 nonzero digits and thus

$$\mathbf{n}(A_1) = 1 \cdot 9 \cdot 9 \cdot 9 = 729$$

Similarly

$$\mathbf{n}(A_2) = \mathbf{n}(A_3) = \mathbf{n}(A_4) = 729$$

and

$$\mathbf{n}(E) = \mathbf{n}(A_1) + \mathbf{n}(A_2) + \mathbf{n}(A_3) + \mathbf{n}(A_4) = 4 \cdot 729 = 2,916.$$

Finally

$$P(E) = \frac{2,916}{10,000} = 0.2916,$$

in other words, approximately 29% of all possible phone numbers in one exchange contain exactly one 0. □

Exercises

1. There are seven canoe routes between an outfitter and a choice fishing spot and only three of them require a portage. How many different ways can a fisherman canoe from the outfiter's to the fishing spot and back without a portage on the way out?

2. How many different breakfasts consisting of juice, cereal, and eggs are available in a cafeteria that serves 5 kinds of juice, 6 kinds of cereal, and 3 kinds of eggs?

3. A bicycle manufacturer offers customers a choice of six different colors, a five or ten speed gear shift, and 4 different frame sizes. From how many different bicycles does a customer have to choose?

4. A travel agent has been asked to organize an overnight bus trip to Gettysburg for a group. The travel agent deals with 3 bus companies, 10 motels in Gettysburg, and 2 battlefield tour guides. How many different ways can the travel agent organize this trip?

5. A quiz consists of four multiple choice questions and each question has three possible answers of which only one is correct. As in Sample Problem 4 an unprepared student decides to guess and with the student's method of guessing all possible answer sheets are equally likely. Compute the probabilities of the following events:

   (a) The student answers the questions all incorrectly,

   (b) the student answers at least one question correctly,

   (c) the student answers exactly one question correctly,

   (d) the student answers exactly one question incorrectly,

   (e) the student answers at least one question incorrectly.

6. As in Sample Problem 6 suppose a phone number is selected at random from one telephone exchange of $10,000$ numbers. What is the probability that the last 4 digits of the number selected will all be less than 4? What is the probability that exactly 3 of the last four digits will be greater than 6?

7. A course name consists of 4 letters followed by 3 digits, e.g. Math 111. Digits may be repeated, but the first digit must be chosen from $0, 1, 2, 3, 4$, and the second and third digit from $0, 1, 2, \ldots, 9$. All 26 letters of the alphabet nay be used. How many course names can be formed?

8. An ornithologist uses bird bands of five different colors to identify individual birds. If she puts two bands of the same size on the right leg, how many birds can she band differently? If she puts two bands on the right leg and one on the left leg, how many birds can she band differently? Same question for two bands on one leg and one band on the other leg.

9. A sophomore meets the prerequisites for 12 English, 10 History, 5 Mathematics and 3 Chemistry courses. Suppose he decides to take one course in each of these subjects. How many different programs are available to this student? How many programs are available to him if he decides to take one course each in three of these four subjects?

10. How many ways can the grades A, B, and C be assigned to a class of 7 students? If the grades are assigned randomly, what is the probability that none of the students will receive an A?

11. A developer is building a community of single family homes. Each home will have a brick and aluminum siding exterior. The architect has produced eleven different exterior designs which are acceptable to the developer. Moreover, the developer has selected two colors of brick and five colors of siding. How many homes with different outside appearances can the developer build?

12. Mr. Jones wants to fly from Washington, D.C. to Los Angeles and return with no more than two stop-overs in the whole trip. Possible stop- over cities are Chicago, Denver and Kansas City. No more than one stop-over can be made in each direction, and stop-over cities can not be repeated. If there are also non-stop flights from Washington to Los Angeles and back, in how many ways can Mr. Jones make the round trip?

13. How many 3 letter "words" begin with a consonant followed by 2 vowels? How many 3 letter " words" contain exactly one consonant? How many 3 letter "words" begin with a consonant and end with a vowel?

14. How many different identification codes can a company generate for its employees by using two letters followed by a two digit number, with both digits selected from 1, 2, 3, 4, 5?

15. In an oral assessment for first graders the children answer each question by using either a green or a red crayon and drawing a square, circle or triangle on a piece of paper. If the children answer 10 questions in this fashion, how many possible answer sheets are there?

16. On a stage there are 4 chairs in a row for 4 dignitaries. How many different ways can the dignitaries be seated on these chairs?

17. The names of fraternities and sororities usually consist of 2 or 3 Greek letters. How many such names are possible? (There are 24 letters in the Greek alphabet.) In how many of these names is no Greek letter used more than once?

18. Three distinguishable dice are rolled.

    (a) What is the probability of not getting a 1 on any of the dice?
    (b) What is the probability of getting at least one 1?
    (c) What is the probability of getting exactly two 6's, or all three equal?

19. An acronym is a word formed from the initial letters of a compound word or title; e.g., NASA, which stands for National Aeronautics and Space Administration. How many acronyms could be made using 3, 4 or 5 letters from the English alphabet?

20. John has 6 pairs of pants, 2 of which are brown and 4 of which are green. He has 10 shirts, 2 of which are white, 3 are blue, and 5 are yellow. He has 4 sweaters, 3 of which are red and 1 is green. He picks at random 1 shirt, 1 sweater, and 1 pair of pants. What

is the probability he picks a yellow shirt, a red sweater, and a green pair of pants? What is the probability he picks a pair of brown pants, a white shirt, and a green sweater? What is the probability he picks a blue shirt or a red sweater?

2.2 Permutations

Order may or may not be a crucial factor when a number of different objects are being selected from a set. Although the same three letters from the alphabet are used to spell "god" and "dog", the words are very different because the letters are written in different orders. However, inviting Ann, Carl, Joan, and Bill to dinner on a Saturday evening is the same as inviting Joan, Ann, Bill, and Carl. Terms like "words", "schedules", and "arrangements" connote order, while terms like "committees" and "hands" (as in cards) suggest an absence of order. Deciding whether or not order is important in a counting problem is often a critical first step in the solution.

Suppose one prize is to be given to each of three different people selected from a group of thirty people by drawing names from a hat. In this context compare the following counting problems:

> PROBLEM A. If there is a first prize of \$150, a second prize of \$100, and a third prize of \$50, how many different results of the drawing are possible?
>
> PROBLEM B. If there are three equal prizes of \$100 each, how many different results of the drawing are possible?

The two problems are certainly very similar. In both, a total of \$300 is to be divided among three winners from the thirty people. A less obvious but more important similarity is that one person may not win more than one prize. In other words, after a name is drawn it may not be replaced or put back in the hat. This is often referred to as selection without replacement. There is also a very fundamental difference. In the first problem the order in which the names are selected is very important. Which prize would you rather win, first or third? Because the prizes are the same in the second problem, the order in which the names are selected can be ignored. In fact the three names could be selected simultaneously. The distinction between these two problems is precisely the distinction between permutations and combinations. In both permutations and combinations a specified number of different objects is taken from a given set. In a permutation the objects are arranged in a definite order, while order plays no role whatsoever in a combination. This section is devoted to permutations and the next to combinations.

Any arrangements of r distinct objects selected from a set of n objects is called a *permutation of n objects taken r at a time.*

EXAMPLE 1. All the permutations of the 3 letters a, b, and c taken 2 at a time are

$$\begin{array}{ccc} ab & ba & ca \\ ac & bc & cb. \end{array}$$

Let $\mathbf{p}(n,r)$ denote the total number of all possible permutations of n objects taken r at time. The immediate problem is to find a simple convenient formula for $\mathbf{p}(n,r)$. This will enable us to calculate the number of permutations of a given type without listing them.

The operation of constructing the permutations of n objects taken r at a time can be broken into r steps - pick the first element, pick the second element, ..., pick the rth or last element of the permutation. This suggests applying the Multiplication Principle to compute $\mathbf{p}(n,r)$. Selecting the first element in the permutation can be done in n ways, i.e., $k_1 = n$. There are now $n-1$ objects left from which the remaining objects in the permutation must be selected. (Note the definition of a permutation requires the arranged objects to be distinct, hence there are no repeats in a permutation.) Consequently, the second element of the permutation can be selected in $n-1$ ways and $k_2 = n-1$. This leaves $n-2$ objects from which the third object must be selected, so $k_3 = n-2$. Repeating the argument as often as necessary we eventually come to the last or r^{th} step which can be done in $n-r+1 = k_r$ ways. By the Multiplication Principle

$$\mathbf{p}(n,r) = n \cdot (n-1) \cdot (n-2) \cdot \ldots \cdot (n-r+1).$$

The above formula can be written more compactly using factorials. The *factorial* of a positive integer n is defined to be the product of the first n consecutive integers. The standard notation for the factorial of n is $n!$ which is read n-factorial. For technical reasons $0!$ is defined to be 1. The first few factorials are:

$$\begin{aligned}
0! &= 1\\
1! &= 1\\
2! &= 1\cdot 2 = 2\\
3! &= 1\cdot 2\cdot 3 = 6\\
4! &= 1\cdot 2\cdot 3\cdot 4 = 24\\
5! &= 1\cdot 2\cdot 3\cdot 4\cdot 5 = 120\\
6! &= 1\cdot 2\cdot 3\cdot 4\cdot 5\cdot 6 = 720\\
7! &= 1\cdot 2\cdot 3\cdot 4\cdot 5\cdot 6\cdot 7 = 5,040\\
8! &= 1\cdot 2\cdot 3\cdot 4\cdot 5\cdot 6\cdot 7\cdot 8 = 40,320\\
9! &= 1\cdot 2\cdot 3\cdot 4\cdot 5\cdot 6\cdot 7\cdot 8\cdot 9 = 362,880\\
10! &= 1\cdot 2\cdot 3\cdot 4\cdot 5\cdot 6\cdot 7\cdot 8\cdot 9\cdot 10 = 3,628,800
\end{aligned}$$

and the general formula is

$$\boxed{n! = 1\cdot 2\cdot 3\cdot \ldots \cdot (n-1)\cdot n.}$$

The trick to obtaining a more compact formula for $\mathbf{p}(n,r)$ is to use cancellation to simplify the following expression:

$$\begin{aligned}
\frac{n!}{(n-r)!} &= \frac{1\cdot 2\cdot 3\cdot \ldots \cdot (n-r)\cdot \ldots \cdot (n-1)\cdot n}{1\cdot 2\cdot 3\cdot \ldots \cdot (n-r)}\\
&= (n-r+1)\cdot \ldots \cdot (n-1)\cdot n\\
&= \mathbf{p}(n,r).
\end{aligned}$$

Therefore,

$$\boxed{\mathbf{p}(n,r) = \frac{n!}{(n-r)!}.}$$

Sample Problem 1. *Calculate* $\mathbf{p}(12,3)$.

SOLUTION: Here $n = 12$ and $r = 3$. Using $\mathbf{p}(n,r) = n!/(n-r)!$ gives us

$$\begin{aligned}\mathbf{p}(12,3) &= \frac{12!}{(12-3)!} = \frac{12!}{9!} \\ &= \frac{1\cdot 2\cdot 3\cdot 4\cdot 5\cdot 6\cdot 7\cdot 8\cdot 9\cdot 10\cdot 11\cdot 12}{1\cdot 2\cdot 3\cdot 4\cdot 5\cdot 6\cdot 7\cdot 8\cdot 9} \\ &= 10\cdot 11\cdot 12 \text{ (by cancellation)} \\ &= 1{,}320.\end{aligned}$$

The cancellation that we used here is just a specific case of the cancellation used to show that $\mathbf{p}(n,r) = n!/(n-r)!$. Moreover, taking advantage of this cancellation is far superior to using long division. □

Besides using the formula to calculate $\mathbf{p}(n,r)$ most calculators have a built in permutation program. Permutations are usually second functions and denoted by ${}_nP_r$ instead of $\mathbf{p}(n,r)$. (The notation ${}_nP_r$ is frequently used for permutations but has the disadvantage that the important numbers n and r are subscripts.) Using a calculator to obtain numerical answers to problems and to gain a sense of the magnitude of numbers involved will be helpful.

EXAMPLE 2. To find $\mathbf{p}(15,8) = {}_{15}P_8$ using a typical calculator you would enter 15, the permutation keystrokes followed by the 8 and the equals to get 259,459,200.

If we return to the problem of determining how many ways can a first, second, and third prize be awarded to three different people from a group of thirty people (Problem A from the opening paragraphs of the section), we see that this amounts to computing the number of permutations of 30 objects taken 3 at a time or $\mathbf{p}(30,3)$. Moreover, we now know how to calculate $\mathbf{p}(30,3)$:

$$\mathbf{p}(30,3) = \frac{30!}{(30-3)!} = \frac{30!}{27!}$$

$$= \quad 28 \cdot 29 \cdot 30 = 24,360.$$

We can not solve Problem B yet, but will return to it in the next section and compare the answers to both problems.

How many permutations of n objects taken n at a time are there? In other words, how many ways can we arrange all the objects? To answer this we must compute $\mathbf{p}(n,n)$. From the formula for $\mathbf{p}(n,r)$ we get

$$\mathbf{p}(n,n) = \frac{n!}{(n-n)!} = \frac{n!}{0!} = \frac{n!}{1} = n!$$

or simply

$$\mathbf{p}(n,n) = n!.$$

From our earlier calculations of factorials it follows that there are over three million different ways in which 10 objects can be ordered.

Sample Problem 2. *Suppose a set of signal flags consists of one red, one blue, one yellow, and one green flag, and signals are sent by flying one or more of these flags on a vertical pole in a specified order. For example, red above yellow might be a distress signal while yellow above red might signal the approach of a storm. How many different signals of this type are possible?*

SOLUTION: First, all the signals must be partitioned into four subsets according to how many flags are used in the signal. Determining the number of signals using a fixed number of flags is a permutation problem, e.g., the number of signals using exactly two flags is $\mathbf{p}(4,2)$. By the Partition Principle the answer is

$$\mathbf{p}(4,1) + \mathbf{p}(4,2) + \mathbf{p}(4,3) + \mathbf{p}(4,4)$$

To get a numerical answer we calculate each term as follows:

$$\mathbf{p}(4,1) = \frac{4!}{(4-1)!} = \frac{4!}{3!} = 4.$$

Note that this agrees with the obvious answer in this case.

$$\mathbf{p}(4,2) = \frac{4!}{(4-2)!} = \frac{4!}{2!} = 3 \cdot 4 = 12$$

$$\mathbf{p}(4,3) = \frac{4!}{(4-3)!} = \frac{4!}{1!} = 4! = 12$$

$$\mathbf{p}(4,4) = 4! = 24.$$

Adding the results of these calculations gives

$$4 + 12 + 24 + 24 = 64$$

and completes the solution. □

Sample Problem 3. *How many "words" can be formed by using 4 different letters from the letters in the word POSTCARD? (The quotes around words in the beginning of the sentence indicate that we do not care whether or not they have any meaning in English or another language.) If one of these words is selected at random, what is the probability that it will not contain any vowels?*

SOLUTION: Since the order of letters in a word with or without meaning is important, this is a permutation problem. Because POSTCARD contains eight distinct letters, the "words" we want to count are all permutations of 8 objects taken 4 at time. By the formula for $\mathbf{p}(n, r)$

$$\mathbf{p}(8,4) = \frac{8!}{(8-4)!} = \frac{8!}{4!} = 5 \cdot 6 \cdot 7 \cdot 8 = 1,680$$

which answers the first question.

The sample space S for the second question is obviously the set of "words" we just counted, and thus $\mathbf{n}(S) = 1,680$. Let E be the event the "word" selected at random does not contain a vowel.

In other words, E is the set of "words" which can be formed by using 4 different letters from the letters PSTCRD. Consequently,

$$\mathbf{n}(E) = \mathbf{p}(6,4) = \frac{6!}{(6-4)!} = \frac{6!}{2!} = 3 \cdot 4 \cdot 5 \cdot 6 = 360$$

and

$$P(E) = \frac{360}{1,680} = .214$$

answers the second question. □

Exercises

1. Evaluate the following:

   (a) $\mathbf{p}(5, 5)$,

   (b) $\mathbf{p}(7, 4)$,

   (c) $\mathbf{p}(40, 1)$,

   (d) $\mathbf{p}(25, 2)$.

2. A TV station manager has 10 commercials she could show during a station break, but federal regulations only allow her to show 3. How many different ways could she schedule 3 of these 10 commercials for a station break?

3. How many 3 digit numbers can be formed using 3 different digits from 1, 2, 3, 4, 5 and 6? How many of these numbers are divisible by 5? How many of these numbers are less than 300?

4. If the letters in the word FLORIDA are randomly rearranged, what is the probability of each of the following:

   (a) The vowels are all together at the beginning,

   (b) the vowels are all together?

5. Six married couples go to a concert. In how many ways can they be seated in 12 chairs in a row? In how many ways can they be seated in 12 chairs in a row such that each person is seated next to his or her spouse?

6. A written test for a driver's license is constructed by arranging 10 multiple choice questions taken from the 40 questions in the study manual. How many different tests are possible?

7. A tourist plans to visit 8 different public buildings and 4 different monuments in Washington, D.C.

   (a) How many different ways can he order these things he wants to do?

(b) How many ways can he do this if he visits the monuments first?

(c) Because he overslept he has time to do only three quarters of these things. With this restriction how many arrangements are possible?

8. A team of policemen is assigned to operate a radar unit Monday, Wednesday, and Friday at a different site each of these days. The sites are chosen each week from 11 suitable sites in their district. How many different weekly schedules are possible?

9. A student wants to arrange her 8 different text books on a shelf. In how many ways can this be done? In how many ways can this be done so that her English and her Mathematics text books are separated by exactly one other text?

10. The first of two questions on a questionnaire is to select your first, second and third choice for president from a list of 10 prominent Democrats. The second question is to do the same thing from a list of 8 prominent Republicans.

    (a) How many different ways can the first question be answered?

    (b) How many different ways can the entire questionnaire be completed?

11. A classroom has 15 seats. The front row has 5 seats, the second row has 6 seats and the third row has 4 seats. Nine students come to sit in the classroom.

    (a) In how many ways can they be seated?

    (b) If 3 of the students are nearsighted and must sit in the front row, and the others can sit anywhere, in how many ways can they be seated?

12. A nursery school classroom has 5 girls and 5 boys. Ten cots are lined up for them to nap on. How many possible sleeping arrangements are possible if

    (a) there are no restrictions,

(b) Erich and Tommy giggle too much and so must be kept separated,

(c) if the oldest 2 children must occupy the first and last cots?

13. A security lock requires two different keys, which must be used in the correct order. If the two required keys are mixed with nine others, what is the maximum number of attempts necessary to open the lock by systematically searching through the eleven keys?

14. A painter has 10 pictures for sale, and wants to display then in one row on a long wall. How many ways can he arrange them? Because two of them are variations of the same scene, he decides they should be separated by at least two other paintings. Now how many ways can he arrange all 10 on the wall?

15. A vehicle license plate in a certain state consists of two letters followed by 3 digits, e.g. AB211. A license plate is randomly selected from all the possible ones for this state. Compute the probability of the following events:

    (a) No letter or digit is repeated on the license plate selected,

    (b) There is at least one vowel on the license plate selected,

    (c) The digits are all the same on the license plate selected.

16. How many different Bingo cards of the standard type are possible? (A standard Bingo card uses 5 different numbers from 1 to 15 in the first column, 5 different numbers from 16 to 30 in the second column, etc. except for the third or middle column which contains four numbers and a free space.) If a Bingo card is selected at random, what is the probability that the numbers in the first column are all less than or equal to 10? If a Bingo card is selected at random, what is the probability that the unlucky number 13 will appear on it?

17. An urn contains six balls numbered 1 to 6. One after another three balls are drawn from the urn and put aside. What is the

probability that the number on the first ball is less than the number on the second ball which in turn is less than the number on the third ball?

18. A set of 26 encyclopedias are lettered A through Z and arranged on a shelf. How many such arrangements are not alphabetical from left to right?

19. Six people are to be seated in a row of six seats.

    (a) In how many ways can this be arranged?

    (b) Suppose two of them - say Romeo and Juliet - are to be seated next to one another. How many seating arrangements are possible?

    (c) How many seating arrangements are there in which two others - say Caesar and Brutus - are not seated side by side?

    (d) How many seating arrangements are there in which Romeo and Juliet are seated next to each other and Caesar and Brutus are not seated side by side?

20. Consider the set of digits $\{1, 2, 3, 4, 5, 6, 7\}$.

    (a) If digits cannot be repeated, how many three digit numerals can be formed from this set of digits?

    (b) If digits cannot be repeated how many even three digit numerals can be formed from this set of digits?

    (c) If digits can be repeated, how many five digit numerals greater than $50,000$ can be formed from this set of digits?

    (d) If digits cannot be repeated, how many numerals less than 1000 can be formed from this set of digits?

2.3 Combinations

A *combination of n objects taken r at a time* is a subset of exactly r objects selected from a set of n objects. Notice that there is no mention of order. In a combination there is no first or last element; all elements of a combination should be regarded equally important.

EXAMPLE 1. The combinations of the 3 objects in $\{a, b, c\}$ taken 2 at a time are

$$\{a, b\}, \{a, c\}, \{b, c\}.$$

(Compare this with the list of permutations of $\{a, b, c\}$ taken 2 at a time which appears in Example 1 of the previous section.)

As with permutations we will be primarily interested in the number of combinations rather than the actual combinations. The total number of possible combinations of n objects taken r at a time will be denoted by

$$\binom{n}{r}$$

This is usually read "n choose r". The first task is to derive a formula for the number of combinations of n objects taken r at a time using the Multiplication Principle.

The strategy for determining the number of combinations is to think again about permutations but from a different point of view. A permutation of n objects taken r at a time can be obtained in a two step operation. First, disregarding order, select r objects at once from the n objects; second, arrange the r objects in some order. To use the Multiplication Principle to determine $\mathbf{p}(n, r)$ the number of ways each step can be done must be determined. First, r objects can be selected at once from n objects in $\binom{n}{r}$ ways which is the number we are trying to find. Second, we know from the previous section that r objects can be arranged in $\mathbf{p}(r, r) = r!$ ways. Now by the Multiplication Principle the total number of permutations of n objects taken r at a time is given by

$$\mathbf{p}(n, r) = \binom{n}{r} \cdot r!$$

and multiplying both sides by $1/r!$ we get

$$\frac{1}{r!} \cdot \mathrm{p}(n,r) = \binom{n}{r}.$$

Now recall that $p(n,r) = n!/(n-r)!$ and hence by substitution

$$\frac{1}{r!} \cdot \frac{n!}{(n-r)!} = \binom{n}{r}.$$

or

$$\boxed{\binom{n}{r} = \frac{n!}{r!(n-r)!}.}$$

Sample Problem 1. *Compute* $\binom{12}{4}$.

SOLUTION: The secret is to cancel as much as possible. In fact, because $\binom{12}{4}$ is an integer, every term in the denominator can be disposed of by cancellation. Specifically,

$$\begin{aligned}
\binom{12}{4} &= \frac{12!}{4! \cdot (12-4)!} = \frac{12!}{4! \cdot 8!} \\
&= \frac{9 \cdot 10 \cdot 11 \cdot 12}{4!} = \frac{9 \cdot 10 \cdot 11 \cdot 12}{1 \cdot 2 \cdot 3 \cdot 4} \\
&= \frac{9 \cdot 5 \cdot 11 \cdot 12}{3 \cdot 4} = 9 \cdot 5 \cdot 11 \\
&= 495.
\end{aligned}$$

and by cancellation the calculation is reduced to multiplying 9 times 5 times 11. □

As with permutations many calculators can compute the number of combinations directly. The comparable notation to ${}_nP_r$ for combinations is ${}_nC_r$.

EXAMPLE 2. To calculate $\binom{15}{8} = {}_{15}C_8$ enter 15, the "${}_nC_r$" key(s), 8 and equals to get 6,435. Notice how much smaller this is compared to $p(15,8) = {}_{15}P_8 = 259,459,200$.

There are some special cases of $\binom{n}{r}$ which are worth remembering:

$$\binom{n}{0} = \frac{n!}{0!(n-0)!} = \frac{n!}{n!} = 1$$

$$\binom{n}{n} = \frac{n!}{n!(n-n)!} = \frac{n!}{n!0!} = 1$$

$$\binom{n}{1} = \frac{n!}{1!(n-1)!} = n$$

$$\binom{n}{n-1} = \frac{n!}{(n-1)!1!} = n$$

Notice that the right hand sides of the first and second equations are the same as are the right hand sides of the third and fourth equations. This is not a coincidence, but a consequence of

$$\boxed{\binom{n}{r} = \binom{n}{n-r}.}$$

The last formula holds because each combination of n objects taken r at a time uniquely determines a combination of n objects taken $n-r$ at a time, namely, the $n-r$ objects that are left after the r objects have been removed.

We can now complete our discussion of Problems A and B stated in the beginning of Section 2.2. Recall that A was a permutation problem and the answer was

$$p(30,3) = 24,360.$$

Also recall that Problem B asked the following: How many ways can three prizes of $ 100 each be awarded to different people from a group of 30 people by drawing names? Because the prizes are all the same there is no natural ordering of the winners as there was in Problem A. We now recognize this as a combination problem which asks us to compute

$$\begin{aligned}\binom{30}{3} &= \frac{30!}{3!\cdot(30-3)!}\\ &= \frac{30!}{3!\cdot 27!} = \frac{28\cdot 29\cdot 30}{6}\\ &= 28\cdot 29\cdot 5 = 4,060.\end{aligned}$$

It is not surprising that the answer to Problem A is much larger than the answer to Problem B; after all, each combination of 30 people taken 3 at a time in the answer to Problem B could be arranged in $3! = 6$ ways to obtain the answer to Problem A. It is easy to check that

$$24,360 = 6\cdot 4,060.$$

Sample Problem 2. *A box contains 20 colored ping-pong balls. There are 5 balls each of the following colors: red, green, yellow, and white. The 5 balls of each color are numbered consecutively from one to five. In other words, there is 1 ball of each color with the number two on it, etc.. Four balls are randomly selected at once.*

(a) What is the probability that four balls of the same color are selected?

(b) What is the probability that exactly three of the four balls selected have the same number on them?

(c) What is the probability that exactly 2 of the four balls selected have the same number on them?

(d) What is the probability that the four balls selected will consist of 2 each of 2 different numbers?

SOLUTION: Since the four balls are randomly selected, we can use the classical formula for probability. Since they are selected at once, the outcome is a combination of 20 balls taken 4 at a time.

Hence the sample space S is the set of all combinations of 20 balls taken 4 at a time. Furthermore, the number of elements in S can be calculated as follows:

$$\begin{aligned} \mathbf{n}(S) &= \binom{20}{4} = \frac{20!}{4! \cdot 16!} = \frac{17 \cdot 18 \cdot 19 \cdot 20}{2 \cdot 3 \cdot 4} \\ &= 17 \cdot 3 \cdot 19 \cdot 5 = 4,845. \end{aligned}$$

For convenience we let A, B, C and D denote the events described in parts (a), (b), (c), and (d). To calculate the probabilities of these events we must first determine $\mathbf{n}(A)$, $\mathbf{n}(B)$, $\mathbf{n}(C)$, and $\mathbf{n}(D)$.

(a) To calculate $\mathbf{n}(A)$ first select one of the four colors and then four of the five balls of that color. Thus

$$\mathbf{n}(A) = \binom{4}{1} \cdot \binom{5}{4} = 4 \cdot 5 = 20$$

and

$$P(A) = \frac{20}{4,845} = 0.0041$$

which is a rather unlikely event.

(b) The number $\mathbf{n}(B)$ is calculated by selecting one of the five numbers and then selecting three balls with that number on them all of which can be done in

$$\binom{5}{1} \cdot \binom{4}{3}$$

ways. One way to select the fourth ball is to select one of the remaining four numbers and one ball with that number on it which can be done in

$$\binom{4}{1} \cdot \binom{4}{1}$$

ways. By the Multiplication Principle

$$\begin{aligned} \mathbf{n}(B) &= \binom{5}{1} \cdot \binom{4}{3} \cdot \binom{4}{1} \cdot \binom{4}{1} \\ &= 5 \cdot 4 \cdot 4 \cdot 4 \cdot = 320 \end{aligned}$$

and hence

$$P(B) = \frac{320}{4,845} = 0.0660$$

(c) Although the stated question here is very similar to that of the previous part, there is an important difference in the solutions. To calculate $\mathbf{n}(C)$ we first select one of the five numbers and then two balls with that number on them as we did in the previous part. This can be done in

$$\binom{5}{1} \cdot \binom{4}{2}$$

ways.

The numbers on the remaining two balls must be different from each other as well as different from the number on the first two balls. To do this we first select two of the remaining four numbers and then one ball of each of these numbers. This can be done in

$$\binom{4}{2} \cdot \binom{4}{1} \cdot \binom{4}{1}$$

ways. As part of the solution it is important to understand why these two numbers must be selected simultaneously using $\binom{4}{2}$. A very common but *incorrect* way of selecting the last two balls is to select a number, a ball with that number on it, another number, and a ball with that number on it obtaining

$$\binom{4}{1} \cdot \binom{4}{1} \cdot \binom{3}{1} \cdot \binom{4}{1}.$$

It is wrong because it distinguishes between the first number selected and the second number selected. In particular, this would count

blue 1, red 4

and

red 4, blue 1

as distinct ways of selecting the last two balls according as 1 is selected first and then 4 or vice versa. Because the two different numbers are each on one of the balls selected there is no way to distinguish them. Where as in (b) the two numbers could be distinguished because one appeared on three balls and the other on only one ball. The same issue will arise in part (d)

To finish (c) the number of ways the first two balls can be selected must be multiplied times the correct number of ways for selecting the last two balls to determine $\mathbf{n}(C)$. Consequently,

$$\begin{aligned} \mathbf{n}(C) &= \binom{5}{1} \cdot \binom{4}{2} \cdot \binom{4}{2} \cdot \binom{4}{1} \cdot \binom{4}{1} \\ &= 5 \cdot 6 \cdot 6 \cdot 4 \cdot 4 = 2,880 \end{aligned}$$

and

$$P(C) = \frac{2,880}{4,845} = 0.59$$

which means we can expect the event B to occur roughly 6 out of 10 times.

(d) For D two numbers must be selected simultaneously and then two balls with each of these numbers on them are selected. Thus

$$\mathbf{n}(D) = \binom{5}{2} \cdot \binom{4}{2} \cdot \binom{4}{2} = 10 \cdot 6 \cdot 6 = 360,$$

and

$$P(D) = \frac{360}{4,845} = 0.074$$

If the two numbers were not selected together with a $\binom{5}{2}$ but one after another with a $\binom{5}{1}$ and a $\binom{4}{1}$, then

red 3, green 3, yellow 5, red 5

and

yellow 5, red 5, red 3, green 3

would be counted as distinct outcomes according as 3 is selected first and then 5 or vice versa. Since the problem specifies that the 4 balls are to be selected simultaneously, order plays no role and both of the above should be regarded as the same outcome.

In any counting problem care must be taken to make sure that things are not counted twice. One common mistake was thoroughly discussed in both parts (c) and (d) and making this mistake doing either part would count everything in the events C or D twice and double the final answer. □

Sample Problem 3. *A store receives a shipment of 24 electronic games. The manager tells an employee to randomly select 5 games at once, test them and return the shipment if 2 or more are defective. Suppose unbeknownst to the manager there are 15 defective ones in the shipment.*

(a) What is the probability that exactly 2 of the games tested will not be defective?

(b) What is the probability that the shipment will not be returned?

(c) What is the probability that the shipment will be returned?

SOLUTION: The sample space S is all combinations of 24 objects taken 5 at a time and

$$\mathbf{n}(S) = \binom{24}{5} = 42,504.$$

As usual, let A, B, and C denote the events described in parts (a), (b), and (c).

(a) If exactly 2 of the games selected are not defective, then exactly 3 must be defective. To calculate $\mathbf{n}(A)$ think of separating the games into two piles, the good and the bad. There are 9 in the good pile and 15 in the bad pile. Select 2 from the good and 3 from the bad. By the Multiplication Principle

$$\mathbf{n}(A) = \binom{9}{2} \cdot \binom{15}{3} = 455 \cdot 36 = 16,380$$

and hence

$$P(A) = \frac{\binom{9}{2} \cdot \binom{15}{3}}{\binom{24}{5}} = \frac{16,380}{42,504} = 0.385$$

(b) The shipment will not be returned if the employee does not find any defective games in the five he tests or if he finds exactly one defective game. This provides a natural partition of B. Let B_5 be the event that all 5 of the tested games are good and let B_4 be the event that exactly four of the games tested are good. Then

$$P(B) = P(B_5) + P(B_4).$$

Using the solution of (a) as a model, it is evident that

$$P(B_5) = \frac{\binom{9}{5} \cdot \binom{15}{0}}{\binom{24}{5}} = \frac{126 \cdot 1}{42,504}$$

and

$$P(B_4) = \frac{\binom{9}{4} \cdot \binom{15}{1}}{\binom{24}{5}} = \frac{126 \cdot 15}{42,504} = \frac{1,890}{42,504}$$

Thus

$$\begin{aligned} P(B) &= \frac{126}{42,504} + \frac{1,890}{42,504} \\ &= \frac{2,016}{42,504} = 0.0474. \end{aligned}$$

(c) The easiest way to do this is to note that $C = B'$. So

$$P(C) = 1 - P(B) = 1 - .0474 = 0.9526$$

and with high probability the shipment would be returned.

Notice that the criteria used to decide whether or not to return the shipment would be quite effective in this case. However, for shipments with fewer defective games it would be less effective and maybe not satisfactory. □

Probabilities like those we just computed in Sample Problem 3 are fairly common and are called *hypergeometric*. The essential features of a hypergeometric situation are the following:

- A set of n objects from which r objects are randomly selected at once (i.e., without replacement).
- A partition of the n objects into two distinct kinds which are usually thought of as good and bad.

Let g denote the number of good objects and let b denote the number of bad objects. Thus $g + b = n$. If $k, 0 \leq k \leq r$, of the selected objects are good, then $r - k$ of the objects selected are bad. By the Multiplication Principle, we can select k good objects and $r - k$ bad objects in

$$\binom{g}{k} \cdot \binom{b}{r-k}$$

different ways. Since the natural sample space is the set of all combinations of n objects taken r at a time, the probability that exactly k of the r objects selected will be good is given by the formula

$$\frac{\binom{g}{k} \cdot \binom{b}{r-k}}{\binom{n}{r}}.$$

Exercises

1. Evaluate the following:

   (a) $\binom{10}{6}$

   (b) $\binom{20}{2}$

   (c) $\binom{8}{5}$

   (d) $\binom{27}{1}$

   (e) $\binom{101}{99}$

   (f) $\binom{85}{84}$

2. A fancy set of glasses contains 8 glasses no two of which are the same color. How many ways can you select 3 glasses from this set to put on a tray for guests to use?

3. A teacher with a class of 21 children wants to watch as many of them as possible working problems at the blackboard. The blackboard is only big enough to accommodate 18 students. How many ways can the teacher select 18 of the 21 students to work at the board?

4. A club consists of 4 married couples. A committee of three people is formed. What is the probability that this committee will be all of the same sex?

5. A high school debating club has 15 members - 7 boys and 8 girls. How many ways can the faculty advisor select 3 of them to represent the school in a debate? How many ways can the 3 debators be selected if the advisor wants to avoid selecting both all boys and all girls?

6. How many ways can you select 5 different breakfast cereals in a grocery store that carries 30 different kinds of cereal? If 10 of the 30 kinds of cereal are presweetened, how many ways can you select the 5 different kinds so that at most 2 are presweetened?

7. Twin sisters, Kori and Paige, have 11 sweaters between them (all different). On Sunday Kori selects 5 sweaters at random to wear to school the coming week. How many different wardrobes of 5 sweaters each can Kori select? From the remaining sweaters how many ways can Paige select a different sweater to wear on each school day of the coming week?

8. A McDonald's manager has 6 hamburger chefs, 5 French fry cooks, 10 cashiers, and 4 people to clean up. He needs 2 hamburger chefs, 1 French fry cook, 5 cashiers, and 2 people to clean up for a particular shift. How many ways can he choose these people?

9. At a company picnic two door prizes of equal value are given out by randomly selecting 2 names at once from the names of the 50 employees at the picnic. What is the probability that both winners will come from a group of 8 employees at the picnic who work in the same office?

10. The pool for a jury contains 35 people - 15 men and 20 women.

    (a) How many ways can a jury of 12 people be selected from this pool?

    (b) If the judge insists that the jury contain at least 5 women and at least 5 men, how many ways can a jury of 12 people be selected from this pool?

11. In the context of Sample Problem 2 four balls are randomly selected at once from a box containing 20 colored ping-pong balls. There are 5 balls each of the following colors: red, green, yellow, and white. The 5 balls of each color are numbered consecutively from one to five. What are the probabilities that the 4 balls selected have the following properties:

    (a) One ball of each color?

    (b) Two each of two different colors?

    (c) At least 3 balls with the same number on them?

12. Thirty qualified people apply for eighteen identical state jobs. There are 10 from the western part of the state, 8 from the central part, and 12 from the eastern part.

    (a) How many ways can these jobs be filled if they hire the same number of people from each region?

    (b) How many ways can these jobs be filled if they hire at most one person from the eastern region?

    (c) How many ways can these jobs be filled if they hire at least 2 people from the eastern region?

13. At a school picnic there are 20 first graders, 20 second graders, and 20 third graders. Six of them are selected at random to clean up afterward. What is the probability that the cleanup committee will consist of two children from each class? What is the probability that the cleanup committee will consist of three children from each of two classes? What is the probability that at least 5 of the students selected to clean up will be from the same class?

14 For a county cross country meet each high school is allowed to enter 5 runners. Stoneville High has 17 runners on its cross country team of which 8 are seniors. How many ways can the Stoneville coach select his 5 entries? If there is a rule that he may enter at most 2 seniors, how many ways can he select his 5 entries?

15 On a shelf in a store there are 36 flashlight batteries of which 12 are dead, i.e. there is no electrical charge left in them. Suppose you randomly select 6 of these batteries at once.

    (a) What is the probability that exactly 2 of the batteries you select are dead?

    (b) What is the probability that at least one of the batteries you select is dead?

16. If 5 coins are selected at random from 15 coins - 10 quarters and 5 nickels - what is the probability that the value of the coins selected will be 85 cents? At least 85 cents?

17. How many different "words" can be formed by rearranging the letters in each of the words PINEAPPLE and REMEMBER? (Hint: First select the places for the repeated letters and then arrange the unrepeated letters in the remaining places.) How many arrangements of each will have the E's together?

18. A grocery store is selling grapefruit at 5 for a dollar. Suppose a customer randomly selects 5 at once from a box of 60 grapefruits of which 40 are pink and 20 are white. What is the probability that the customer selects exactly 3 white ones? Same question for 5 of the same kind.

19. To make the first row of a quilt, a woman arranges 18 colored squares of cloth in a row. She has 1 pink, 1 orange, 1 green, 4 yellow, 5 blue, and 6 red squares. How many different ways can she make the first row?

20. A computer program is identified by a "name" consisting of 4 or 5 letters, for example, BAKQ or XYXAZ.

    (a) How many names are possible?

    (b) In how many of the 4-letter names is no letter repeated?

    (c) How many of the 5-letter names can be formed using 2 A's and 1 each of the letters B, C, and D?

    (d) How many of the 4-letter names contain the letter Z at least once?

21. How many numbers can be formed by rearranging the digits 37726387?

22. A book store has 16 different books on sale, 9 fiction and 7 non-fiction. How many different ways could you purchase 6 of these books so that your purchase would include at least two fiction and at least 2 non- fiction?

23. A farmer's dairy cooperative receives a shipment of 20 cans of milk from a member farmer and tests 6 of them. If one or more cans fail the test, the shipment is rejected; if four or more fail, the

farmer is suspended from the co-op for a month. Suppose 8 cans in this shipment would not pass. What is the probability that the shipment will be rejected but the farmer will not be suspended?

24. A baseball manager has 4 pitchers, 7 infielders, 3 catchers, and 10 outfielders available for a game. How many ways can he select a pitcher, 4 infielders, a catcher, and 3 outfielders to start the game? (After he selects a player in a category he will assign him to that type of position.)

25. A certain set of alphabet blocks consists of 26 red blocks, 26 blue blocks, 26 yellow blocks, and 26 green blocks. Each of the 26 letters in the English alphabet appears on one block of each color. The blocks are placed in a large basket and 6 of them are randomly selected at once.

    (a) What is the probability that all 6 of the blocks selected will be of the same color?

    (b) What is the probability that the 6 blocks selected will include 4 blocks with the same letter on them?

    (c) What is the probability that the 6 blocks selected will consist of 3 each of 2 different letters?

26. A store has a special on socks - buy 6 pairs and get 1 pair free. Suppose you randomly select 7 pairs from a bin containing 40 pairs without knowing that 25 of these pairs of socks are of a decidedly inferior quality.

    (a) What is the probability that exactly 4 of the pairs of socks you select will be inferior?

    (b) What is the probability that not all of the pairs of socks you select will be inferior?

27. How many different 5 person committees could be formed from the 100 U. S. Senators?

28. To open a security lock on a door you must push one of five buttons numbered from 1 to 5, then push two of them at the same

time, and push a final button by itself. How many combinations are possible for this lock? How many combinations are possible if pushing two at the same time can be the first, second, or third step instead of only the second step? How many combinations are possible if each button can be used at most once?

Card games are a source of a great many counting and probability problems which can be solved using combinations. The next five problems are based on familiar card games which are played with a standard deck of cards. A standard deck contains 52 cards which are divided equally into 4 suits - clubs (black), diamonds (red), hearts (red), and spades (black). There are 13 cards in each suit and half of the deck is red (26 cards) and the other half is black. Each suit contains an ace, king, queen, and jack; the remaining 9 cards are numbered consecutively from 2 to 10. Thus in each suit there are 13 different kinds of cards or denominations. Kings, queens and jacks are called face cards because of the pictures on them.

29. Many games start by drawing a card to see who deals first. If you select a card at random from a standard deck, what is the probability that the card you select is an ace? A face card? A card numbered from 2 to 9?

30. If you are randomly dealt two cards from a standard deck, what is the probability of getting an ace and a face card? (An ace and a face card is called blackjack in that game.) Two face cards?

31. Blackjack hands frequently contain three cards from a standard deck. Compute the number of blackjack hands which have the following properties:

    (a) Two cards, each worth 10, (i.e. K, Q, J, or 10) and an ace.

    (b) At least one 5, and the total value of the hand is 21. (An ace is worth 1 or 11.)

32. When play begins a cribbage hand contains 4 cards from a standard deck. How many cribbage hands contain:

    (a) Two face cards (K, Q, J) and two 5's?

(b) a three card sequence and the fourth card of the same denomination as the middle card of the sequence; e.g. 6 of diamonds, 7 of hearts and clubs, and 8 of diamonds? (In cribbage the cards in a sequence do not have to be in the same suit and the cards are ordered as follows: A, 2, 3, 4, 5, 6, 7, 8, 9, 10, J, Q, K.)

33. What is the probability of being dealt (randomly selected at once) a 5 card poker hand of each of the following standard types:

(a) Royal flush (A, K, Q, J, 10 of the same suit),

(b) straight flush (5 cards of the same suit in sequence in which an ace may be a "one", but not a royal flush),

(c) four of a kind; e.g. 4 kings,

(d) full house (3 of a kind and a pair; such as 3 sevens and 2 jacks),

(e) flush (five cards from the same suit but not in sequence),

(f) straight (any 5 cards in sequence but not from the same suit),

(g) three of a kind (exactly 3 cards of the same kind and 2 cards of different denominations),

(h) two pairs (this excludes 3 or 4 of a kind),

(i) one pair (exactly 2 cards of the same kind and the other 3 cards unmatched)?

34. A gin rummy hand contains 10 cards from a standard deck of 52 cards. How many gin rummy hands have each of the following properties:

(a) Exactly three face cards,

(b) a five card sequence in one suit, a 3 card sequence in another suit, and no other cards in these two suits. (An ace can only be a 1),

(c) three each of two denominations and four of a third denomination?

35. What is the probability of being dealt a bridge hand without any honors? (A bridge hand consists of 13 cards from a standard deck and an honor is an ace, king, queen, jack, or 10). What is the probability of being dealt a bridge hand containing 7 cards from one suit and 2 each from the other 3 suits?

2.4 The Binomial Theorem

Combinations also have an important use in elementary algebra. Recall that an expression containing two terms like $a+b$ is called a binomial, and

$$(a+b)^2 = a^2 + 2ab + b^2.$$

If we multiply the right hand side by $a+b$ and collect like terms, we get

$$(a+b)^3 = a^3 + 3a^2b + 3ab^2 + b^3.$$

Similarly

$$(a+b)^4 = a^4 + 4a^3b + 6a^2b^2 + 4ab^3 + b^4$$

and so forth, but this is a tedious process. So we might ask for a better way to compute $(a+b)^n$ and this is where combinations are useful.

Consider $(a+b)^5$ or $(a+b)(a+b)(a+b)(a+b)(a+b)$. This product is obtained by picking either the a or b from each of the five binomials in parentheses, multiplying them together, and then adding all such terms. We will systematically count the number of terms in which b appears to each fixed power from 0 to 5.

If we select the a from each of the binomials, we get simply

$$a^5 = a^5b^0$$

(Remember $x^0 = 1$ for any x). If we select one of the five binomials, use the b from it, and use the a's from the others, we get a^4b. Since this one binomial can be selected in $\binom{5}{1}$ ways, the next term in $(a+b)^5$ is

$$\binom{5}{1}a^4b.$$

Now select two of the five binomials, use the b's from them, and use the a's in the others. Doing this in all possible ways produces the term

$$\binom{5}{2}a^3b^2.$$

Similarly one obtains the terms

$$\binom{5}{3}a^2b^3 \text{ and } \binom{5}{4}ab^4.$$

Of course, selecting all the b's gives

$$b^5.$$

Since this accounts for all possible terms in $(a+b)^5$, it follows that

$$(a+b)^5 =$$

$$a^5 + \binom{5}{1}a^4b + \binom{5}{2}a^3b^2 + \binom{5}{3}a^2b^3 + \binom{5}{4}ab^4 + b^5 =$$

$$\binom{5}{0}a^5 + \binom{5}{1}a^4b + \binom{5}{2}a^3b^2 + \binom{5}{3}a^2b^3 + \binom{5}{4}ab^4 + \binom{5}{5}b^5$$

because $\binom{5}{0} = 1 = \binom{5}{5}$. (There is one way to not choose any b's and one way to choose all b's.)

From this analysis of $(a+b)^5$ it is natural to infer that

$$\begin{aligned}(a+b)^n &= \binom{n}{0}a^n + \binom{n}{1}a^{n-1}b + \binom{n}{2}a^{n-2}b^2 + \ldots + \\ &\binom{n}{n-2}a^2b^{n-2} + \binom{n}{n-1}ab^{n-1} + \binom{n}{n}b^n\end{aligned}$$

which is called the *Binomial Theorem.* Notice that with the Binomial Theorem we can calculate $(a+b)^n$ without multiplying any binomials; all we need do is determine the coefficients using $\binom{n}{r} = \frac{n!}{r!(n-r)!}$ or a calculator. Because of their importance in the Binomial Theorem, the numbers $\binom{n}{r}$ are often called *binomial coefficients.*

Sample Problem 1. *Expand* $(2x-y)^6$.

SOLUTION: We will use the Binomial Theorem with $n = 6, a = 2x,$ and $b = -y$. By substituting these choices in the Binomial Theorem we obtain

$$(2x-y)^6 = \binom{6}{0}(2x)^6 + \binom{6}{1}(2x)^5(-y)^1$$

$$\begin{aligned}
&+ \binom{6}{2}(2x)^4(-y)^2 + \binom{6}{3}(2x)^3(-y)^3 \\
&+ \binom{6}{4}(2x)^2(-y)^4 + \binom{6}{5}(2x)(-y)^5 + \binom{6}{6}(-y)^6 \\
&= \binom{6}{0}2^6x^6 - \binom{6}{1}2^5x^5y + \binom{6}{2}2^4x^4y^2 \\
&- \binom{6}{3}2^3x^3y^3 + \binom{6}{4}2^2x^2y^4 - \binom{6}{5}2xy^5 + \binom{6}{6}y^6.
\end{aligned}$$

To simplify this further we must calculate the binomial coefficients for 6, and some powers of 2.

$$\begin{aligned}
\binom{6}{0} &= \binom{6}{6} = 1 \\
\binom{6}{1} &= \binom{6}{5} = 6 \\
\binom{6}{2} &= \binom{6}{4} = \frac{6!}{2! \cdot 4!} = 15 \\
\binom{6}{3} &= \frac{6!}{3! \cdot 3!} = 20
\end{aligned}$$

$$2^2 = 4, 2^3 = 8, 2^4 = 16, 2^5 = 32, 2^6 = 64.$$

By substituting these calculations

$$(2x - y)^6 = \\ 64x^6 - 192x^5y + 240x^4y^2 - 160x^3y^3 + 60x^2y^4 - 12xy^5 + y^6.$$

is the answer. □

How many subsets are in a set of n elements? Surprisingly a simple solution to this problem can be found with the Binomial Theorem.

The first step is to substitute 1 for both a and b and simplify. When $a = b = 1$, the left hand side of the Binomial Theorem becomes

$$(1+1)^n = 2^n$$

and the right hand side becomes

$$\binom{n}{0} + \binom{n}{1} + \binom{n}{2} + \ldots + \binom{n}{n-1} + \binom{n}{n}$$

because $1^r = 1 = 1^{n-r}$. Hence

$$2^n = \binom{n}{0} + \binom{n}{1} + \binom{n}{2} + \ldots + \binom{n}{n-1} + \binom{n}{n}$$

Now a careful interpretation of the right hand side of this formula using the Partition Principle will answer the question. The collection of all subsets of a set on n objects can be partitioned according to the number of objects in the subset. Since a subset containing r objects selected from a set of n objects is just a combination of n objects taken r at a time , the number of subsets containing r elements is $\binom{n}{r}$. Then by the Partition Principle the total number of subsets which can be found in a set on n elements is

$$\binom{n}{0} + \binom{n}{1} + \binom{n}{2} + \ldots + \binom{n}{n-1} + \binom{n}{n}$$

which is the right side of the previous formula. Therefore, a set of n elements contains exactly 2^n different subsets.

EXAMPLE 1. The $2^3 = 8$ subsets of $\{a, b, c\}$ are listed below according to this partition:

$\emptyset$	$\binom{3}{0} = 1$
$\{a\}, \{b\}, \{c\}$	$\binom{3}{1} = 3$
$\{a, b\}, \{a, c\}, \{b, c\}$	$\binom{3}{2} = 3$
$\{a, b, c\}$	$\binom{3}{3} = 1$
	Total $= 8$

Sample Problem 2. *After being given a list of 7 symptoms, a medical student is asked which of these symptoms (including all or none) would occur with a particular disease. How many different answers to this question are possible?*

SOLUTION: Any subset of the 7 symptoms is a possible answer. A set of 7 elements contains $2^7 = 128$ subsets which is the number of possible answers. □

Exercises

1. Use the Binomial Theorem to expand $(x+3)^4$ and $(s-t)^7$.

2. Use the Binomial Theroem to expand $(z+2y)^5$.

3. Using the Binomial Theroem find the coefficient of a^3b^6 in $(a+b)^9$, and the coefficient of x^3 in $(3-2x)^6$.

4. Find the coefficients of y^3x^6 in $(3y+x)^9$ and w^9z^3 in $(w-2z)^{12}$.

5. How many subsets of $S = \{a, e, i, o, u\}$ are there?

6. A sorority has before it eight applications for membership at a meeting for the election of new members. There are no restrictions on how many new members they can or must elect. How many different outcomes of this meeting are possible?

7. How many subsets of $\{a, b, c, d, e, f\}$ contain at most five elements?

8. Use the Binomial Theroem to find the coefficient of x^8 in $(x-2)^{10}$.

9. How many ways can you select at least one book from six books you have not read to take with you on a long trip?

10. In addition to lettuce and 4 kinds of dressing, a salad bar contains 10 items. How many different salads containing lettuce, one kind of dressing and at least one other ingredient could be fixed at this salad bar?

11. A busy family of 6 never knows which family members will be home for dinner. How many different subsets of this family containing at least two family members could find themselves together at the dinner table some evening?

12. Find the coefficient of y^3x^5 in $(x+y)^8$.

2.5 Review Problems

1. A cookbook contains 8 salad, 12 meat, 10 vegetable, and 15 dessert recipes. How many different dinners consisting of a salad, a meat dish, a vegetable and a dessert could be made from this cookbook? If the dinner includes 2 different vegetables instead of 1, how many dinners are possible?

2. An urn contains 10 numbered balls - 5 red and 5 green. Both the red and green balls are numbered from 1 to 5. Three balls are randomly selected at once from the urn.

   (a) What is the probability that 3 balls of the same color are selected?

   (b) What is the probability that the numbers on the 3 balls that are selected are all different?

   (c) What is the probability that the 3 balls selected are of the same color and the numbers on them are consecutive integers?

3. A hardware store sells two brands of power lawn mowers. They have in stock one each of 5 different models of brand X mowers and one each of 4 different models of brand Y mowers, but they only have room to display 6 of them in a row along a wall in the store. How many different displays of 6 lawn mowers in a row do they have to choose from? If in this row they want to display 3 brand X mowers next to each other and 3 brand Y mowers next to each other, how many different lawn mower displays are possible?

4. On a table at a roadside produce stand there are 100 ears of fresh corn - 60 yellow and 40 white. Suppose you randomly select 12 at once. Compute the probability that you select the following:

   (a) exactly 4 yellow ears,

   (b) at most one yellow ear,

   (c) at least 5 of each kind,

   (d) at least one yellow ear.

5. A football coach selects 6 simple hand motions to use to signal plays to his quarterback. Each signal will consist of 1, 2, or 3 of the simple hand motions done in sequence.

   (a) How many different signals are possible if the simple hand motions may be repeated in a signal?

   (b) How many different signals are possible if the simple hand motions may not be repeated in a signal?

6. A subset A of U is called a proper subset of U provided $A \neq U$ and $A \neq \emptyset$. How many proper subsets of $U = \{a, e, i, o, u\}$ are there?

7. A craft union holds an election to select 3 executive officers - president, first vice-president, and second vice-president, and a steering committee. There are 7 candidates for executive office and 11 candidates for the steering committee. On the ballot each member must indicate his first, second, and third choice for president from the 7 candidates and then circle 5 names from the list of the 11 candidates for the steering committee printed on the ballot. How many different ways can this ballot be correctly executed? How many different ways can the ballot be completed if giving a third choice for president is optional?

8. If 3 Seven-Up cans, 6 Coca-Cola cans, 4 Dr. Pepper cans, and 7 Sprite cans are randomly arranged in a row on a shelf, what is the probability that all the cans of each of these four brands will be together?

9. In your dance class the instructor taught 15 different movements. For a class presentation you have to perform 4 or 5 movements in sequence. What is the number of different presentations possible if:

   (a) there are no restrictions,

   (b) no movement may be used more than once,

   (c) the first and last movements must be the same,

(d) the first and last movements must be the same and there are no other repeats?

10. A deck of cards for the game of Rook contains 56 colored and numbered cards. There are 14 of each of the colors red, black, green and yellow and the cards of each color are numbered from 1 to 14. Six cards are randomly selected at once from a Rook deck.

    (a) What is the probability that these 6 cards will consist of 3 each of two colors?

    (b) What is the probability that only two different numbers will appear on these 6 cards?

    (c) What is the probability that these 6 cards will consist of 2 each of 2 different numbers and no other duplicate numbers?

11. If five married couples are randomly seated in a row at the theater what is the probability of each of the following:

    (a) the men and women alternate,

    (b) no two men sit beside each other? (Hint: (b) is not the same as (a).)

12. A department store has 45 different shirts of the same size on a shelf; 15 are striped, 15 are plaid, and 15 are solid colors. A customer selects 6 of them at random without replacement. What is the probability that the customer selects 2 of each kind? What is the probability that the customer selects 3 shirts of each of 2 kinds? What is the probability that the customer selects at most one of at least one of the three kinds? (Hint: what is the complement?)

13. During a game of Scrabble you have the letters X, A, Z, E, J, R and V and want to use exactly 5 of them to get a triple word score. From among how many five letter "words" can you search for one with meaning? How many contain at least one vowel?

14. The dairy case in a store contains 30 containers of yogurt. Because of careless handling 12 of them have spoiled. An unsuspecting customer randomly selects 5 at once. What is the probability

that the customer selects exactly 3 containers of spoiled yogurt? What is the probability that the customer selects at most 2 containers of spoiled yogurt?

15. Use the Binomial Theorem to answer the following question:

    (a) What is the coefficient of x^7 in $(x-1)^{12}$?

    (b) What is the coefficient of y^6 in $(z+y^2)^9$?

    (c) What is the coefficient of w^4 in $(w+w^{-3})^{12}$?

    (d) What does $(1.1)^4 = (1+0.1)^4$ equal?

16. A local civic group wants to sponsor a small but good parade. They decide to select 20 units to participate from 35 applicants consisting of 15 bands, 11 floats, and 9 clown groups. They also decide the parade should contain 8 bands, 7 floats, and 5 groups of clowns. It is to be assumed that the units are all distinguishable from each other.

    (a) In how many ways can they select the 20 units as prescribed?

    (b) After the 20 units have been selected, how many ways can the parade be formed?

    (c) After the 20 units have been selected, how many ways can the parade be formed if it must begin and end with a band?

    (d) After the parade is over, in how many ways can the judges award a first, second and third prize to 3 of the 7 participating floats?

17. To start the game of dominoes, you select at random seven dominoes from the set.

    (a) How many different starting sets are there? (A domino is a tile marked into two squares; each square has 0, 1, 2, 3, 4, 5 or 6 dots on it. A set of dominoes has the maximum number of tiles which can be marked differently. A standard set of dominoes contains 28 tiles. Why?)

    (b) What is the probability of selecting a starting set containing the double 5 or the double 6?

(c) How many dominoes are in a set of double twelves, where each square on the tiles can be marked with $0, 1, 2, 3, \ldots, 12$ dots?

18. A four letter "word" is randomly selected from all possible four letter "words". Determine the probability of each of the following events:

    (a) the "word" selected begins with a consonant and ends with a vowel, (Use only a, e, i, o, u as vowels.)

    (b) No letter appears more than once in the "word" selected,

    (c) the selected "word" contains exactly two vowels,

    (d) the selected "word" contains at least two different consonants and no repeated consonants,

    (e) the selected "word" begins or ends with a vowel.

19. An electronically readable label for merchandise consists of eight rectangles. There are two possible lengths and three possible widths for each rectangle. How many different labels are there?

20. If there are three empty tables in a restaurant for 2, 4, and 5 people respectively, how many ways can a party of 11 people split up to sit at these tables?

21. How many different Social Security numbers can be formed by rearranging the digits in your Social Security number?

22. A state legislature is composed of 20 Democrats, 18 Republicans, and 2 Socialists. From this group of 40 people a committee of 7 is picked at random.

    (a) How many committees are possible?

    (b) What is the probability that exactly 2 Republicans are picked?

    (c) What is the probability that at least one Socialist is picked?

    (d) What is the probability that all 7 are from the same party?

    e) What is the probability that at least 6 Republicans are picked?

23. Consider the set of digits $\{1, 3, 4, 5, 7, 8, 9\}$.

    (a) If digits cannot be repeated, how many four digit numbers can be formed from this set of digits?

    (b) If digits cannot be repeated and if the middle digit must be an odd digit, how many three digit numbers can be formed from this set of digits?

    (c) If digits cannot be repeated, how many three digit numbers less than 700 can be formed from this set of digits?

    (d) If digits cannot be repeated, how many two digit numbers can be formed from this set of digits with the restricition that the first digit is odd or the last digit is even?

    e) If digits can be repeated, how many positive numbers less than 1000 can be formed from this set of digits?

24. Suppose three basketball teams of 5 players each are being formed from 15 people. There are three centers who must be on different teams, but there are no other restrictions. How many teams can be formed?

25. A stockbroker recommends eight stocks. Unknown to him or us, five of these will make money, the other three will not. We buy two of the issues at random and decide to continue our association with the stockbroker provided at least one of these turns out to be profitable. What is the probability we will continue with the broker?

26. A group of 40 students consists of 10 freshmen, 10 sophomores, 10 juniors, and 10 seniors. Four students are selected at random.

    (a) In how many ways can the selection be made if all the selected students belong to the same class?

    (b) In how many ways can the selection be made if exactly 2 selected students belong to the same class?

    (c) What is the probability that the four selected students are members of different classes?

27. To start the game of Mastermind you secretly place four colored pegs in four holes in a row and then your opponent tries to deduce the pattern of colored pegs you selected. You are given a large supply (so repeats are possible) of red, green, yellow, blue, black and white pegs.

    (a) In how many different ways can you start this game?

    (b) Suppose you decide not to repeat any color. Now how many patterns are available to you to start the game?

    (c) How many ways can you start the game using 1 red, 1 green, and 2 yellow pegs?

    (d) How many ways can you start the game using at most one black peg?

28. In how many ways can 11 boxes of cereals be arranged on a shelf if 3 are boxes of Cheerios, 4 are Corn Flakes, 2 are Rice Puffs and 2 are Wheat Chex?

29. An ice chest contains 36 cans of cola and 12 cans of root beer. If you select 8 cans randomly at once, compute the probability of the following events:

    (a) Exactly 6 cans of cola are selected,

    (b) At least 3 cans of each flavor are selected.

30. How many 5-letter "words" can be formed (using the 26 letters of the English alphabet) if the first and third letters must be vowels (AEIOU), the second, fourth and fifth must be consonants, and no letter many be used more than once? How many 5-letter "words" can be formed which contain exactly two vowels, repeats being allowed? How many 5-letter "words" can be formed which contain exactly 3 vowels and in which no letter is repeated?

31. A bag contains 45 gum drops - 10 lemon, 15 grape, and 20 licorice. If 6 gum drops are chosen from the bag at random, what is the probability of each of the following events:

    (a) They all have the same flavor.

(b) There are 2 of each flavor.

(c) At least one licorice gum drop is selected.

(d) There are at least 3 grape and 2 lemon gum drops selected.

32. A school legislature has 15 members broken down as follows: 5 seniors, 4 juniors, 4 sophomores, and 2 freshmen. A committee of 7 people is chosen at random from the whole legislature. Find the probabilities of the following events:

    (a) There are exactly 2 juniors on the committee.

    (b) There is at least 1 senior on the committee.

    (c) There are 4 committee members who are either seniors of juniors.

33. Use the Binomial Theorem to expand $(1 - y)^6$.

34. Each phone number in a certain city consists of 7 digits, the first of which must be a 3 or 4.

    (a) How many phone numbers are possible?

    (b) How many possible phone numbers start with 4 and contain the digit 5 at least once?

    (c) How many possible phone numbers can be formed using three 2's and one each of 3, 5, 6 and 7?

    (d) How many possible phone numbers have no repeated digits?

Chapter 3

Conditional Probability

At regular intervals the Labor Department announces the current unemployment statistics. This announcement may state that 7% of the work force is unemployed and 12% of the work force between the ages of 18 and 25 is unemployed. How can this information be interpreted probabilistically? Suppose a person is randomly selected from the work force. The probability of selecting an unemployed person is 0.07, but if it were known that the person selected is between 18 and 25 years old, the probability that the person selected is unemployed is 0.12. Thus the condition that the person selected is between 18 and 25 substantially increases the probability of unemployment. This is what is called a conditional probability. The conditional probability of an event may be greater than, equal to, or less than the original probability of the event. For example, given that the person randomly selected from the work force lives in New York State, the probability that he or she is unemployed might be 0.07, while it might be 0.04 given that the person selected has a college education.

When the probability of an event is not changed by a specific condition, which is really just another event, the situation is of special interest and means the probability of the event is independent of the conditioning event. Sometimes this independence is discovered by calculating the conditional probability. Other times the situation strongly suggests that an event is independent of a condition. By assuming that this independence is true, the calculation of certain probabilities is simplified. In particular, when an experiment is to be repeated it is highly desirable to arrange things so that the second outcome is always independent of the first outcome, etc..

The simplest context for analyzing repetitions of an experiment is a Bernoulli trial. It is a simple two outcome experiment like tossing a coin. By focusing on a single event in a sample space a Bernoulli trial can be extracted from a more complicated phenomenon. For example, trying to roll doubles with a pair of dice is a Bernoulli trial which can be repeated many times. Although it is tempting to believe that after failing to roll doubles for a number of times the probability of succeeding increases, this is not even close to being true and understanding why it is false lies at the heart of calculating the probability of patterns of successes and failures for independent Bernoulli trials. In particular, long streaks of failures have positive probability and thus do occur but with a small relative frequency.

3.1 The Defining Equation

Let A and B be events in a sample space S and suppose $P(B) \neq 0$. Then the *conditional probability of A given B* is defined by

$$\boxed{P(A|B) = \frac{P(A \cap B)}{P(B)}.}$$

In the notation $P(\ |\ )$, the event written to the right of the vertical line is always the condition which has been given, and the expression $P(A|B)$ is read "the probability of A given B".

If we think of the probability of an event as the percentage of the sample space S the event occupies, then the conditional probability is just the percentage of B occupied by the event A . For example, if 15% of all personal income tax forms filed with the IRS contain some incorrect information and 6% of all personal income tax forms filed are fraudulent, then $6/15 = 2/5 = 0.4$ or 40% of the incorrect forms are fraudulent, which is the same as saying the conditional probability of a personal income tax return being fraudulent, given that it is incorrect, is 0.4.

Conditional probabilities can also be thought of as the result of making the sample space smaller. The phrase "given B" means that for some reason we know or are assuming that B has occurred or must occur. Thus B', the part of S outside of B, is no longer relevant, and the sample space S can be replaced with the new sample space B. The formula

$$P(A|B) = \frac{P(A \cap B)}{P(B)}.$$

adjusts the size of the probabilities in the sample space B so that Axioms 1, 2, and 3 hold. Actually Axiom 2 is the crucial one because usually $P(B) \neq 1$, but

$$P(B|B) = \frac{P(B \cap B)}{P(B)} = \frac{P(B)}{P(B)} = 1.$$

Sample Problem 1. *A large state university has gathered the following statistical data about its undergraduate students in a typical* *semester:*

	INSTATE	OUT OF STATE
AT LEAST 15 CREDITS	45%	35%
LESS THAN15 CREDITS	15%	5%

In this table the 45% in the upper left means that each semester instate students who earn 15 credits or more comprise 45% of the undergraduate population.

(a) Given that a random student is an instate student, what is the probability that he or she will earn 15 credits or more in the current semester?

(b) Given that a random student is an out of state student, what is the probability that he or she will earn 15 credits or more in the current semester?

(c) Given that a random student earned 15 credits or more in a certain semester, what is the probability that the student is an instate student?

SOLUTION: Let I denote the event that the random student is an instate student, and let F denote the event that he or she earns 15 credits during the current semester. Then the table gives the probabilities of the events $I \cap F$, $I \cap F'$, etc. Specifically, $P(I \cap F) = 0.45$, $P(I \cap F') = 0.15$, $P(I' \cap F) = 0.35$, and $P(I' \cap F') = 0.05$. Because an instate student is either in F or not in F, the events $I \cap F$ and $I \cap F'$ form a partition of I and $P(I) = 0.45 + 0.35 = 0.8$. Similarly $P(F) = 0.6$. Now the questions can easily be answered using the defining formula for conditional probability.

(a) Given that a random student is an instate student, the probability that he or she will earn 15 credits or more in the current semester is

$$P(F|I) = \frac{P(F \cap I)}{P(I)} = \frac{0.45}{0.6} = 0.75.$$

(b) Given that a random student is an out of state student, the probability that he or she will earn 15 credits or more in the current semester is

$$P(F|I') = \frac{P(F \cap I')}{P(I')} = \frac{0.15}{0.40} = 0.375.$$

Notice how a condition can change a probability - $P(F|I)$ is less than $P(F)$ and $P(F|I')$ is greater than $P(F)$.

(c) Given that a random student earned 15 credits or more in a certain semester, the probability that the student is an instate student is

$$P(I|F) = \frac{P(I \cap F)}{P(F)} = \frac{0.45}{0.8} = 0.5625$$

Notice that $P(F|I)$ need not equal $P(I|F)$.

It is equally simple to calculate the probability that a random student is instate or out of state given that he or she earned less than 15 credits in the current semester. These problems are left to the reader. □

When using the classical formula, $P(E) = \mathbf{n}(E)/\mathbf{n}(S)$, to calculate conditional probabilities, the idea that the sample space is just being made smaller becomes quite evident. Calculating $P(A|B)$ with the classical formula produces

$$P(A|B) = \frac{\mathbf{n}(A \cap B)/\mathbf{n}(S)}{\mathbf{n}(B)/\mathbf{n}(S)} = \frac{\mathbf{n}(A \cap B)}{\mathbf{n}(S)} \cdot \frac{\mathbf{n}(S)}{\mathbf{n}(B)} = \frac{\mathbf{n}(A \cap B)}{\mathbf{n}(B)}$$

or simply

$$\boxed{P(A|B) = \frac{\mathbf{n}(A \cap B)}{\mathbf{n}(B)}}$$

To illustrate conditional probabilities in the classical context we return to rolling a pair of dice. Recall that we had to keep track of the outcome of each die and used the following sample space to accomplish this.

$$S = \left\{ \begin{array}{l} (1,1),(1,2),(1,3),(1,4),(1,5),(1,6), \\ (2,1),(2,2),(2,3),(2,4),(2,5),(2,6), \\ (3,1),(3,2),(3,3),(3,4),(3,5),(3,6), \\ (4,1),(4,2),(4,3),(4,4),(4,5),(4,6), \\ (5,1),(5,2),(5,3),(5,4),(5,5),(5,6), \\ (6,1),(6,2),(6,3),(6,4),(6,5),(6,6) \end{array} \right\}$$

Sample Problem 2. *A pair of dice is rolled. Given that a 6 occurs on a least one die, what is the probability that the sum of their outcomes is more than 9?*

SOLUTION: Let A be the event that the sum of the outcomes is more than 9 and let B be the event that a 6 occurs on at least one die. The problem is to determine $P(A|B)$. Using the above sample space,

$$A = \{(4,6),(5,5),(5,6),(6,4),(6,5),(6,6)\}$$

and

$$B = \left\{ \begin{array}{l} (1,6),(2,6),(3,6),(4,6),(5,6),(6,1), \\ (6,2),(6,3),(6,4),(6,5),(6,6) \end{array} \right\}$$

It follows that

$$\begin{aligned} A \cap B &= \{(4,6),(5,6),(6,4),(6,5),(6,6)\}, \\ \mathrm{n}(A \cap B) &= 5, \\ \mathrm{n}(B) &= 11. \end{aligned}$$

Now $P(A|B)$ is calculated as follows:

$$P(A|B) = \frac{\mathrm{n}(A \cap B)}{\mathrm{n}(B)} = \frac{5}{11}.$$

to finish the solution.

Alternatively, we could have found $P(B) = 11/36$ and $P(A \cap B) = 5/36$ and obtained

$$P(A|B) = \frac{P(A \cap B)}{P(B)} = \frac{\frac{5}{36}}{\frac{11}{36}} = \frac{5}{11}$$

which agrees with the first solution. □

Sample Problem 3. *Conditional probabilities occur in the game of craps. The game begins with the shooter rolling a pair of dice. The shooter wins with 7 or 11 (called a natural), loses with 2, 3, or 12 (called craps), and the remaining possibilities - 4, 5, 6, 8, 9, and 10 - become the shooter's point for the rest of the game. (In this context all the numbers naturally refer to the sum of the outcomes of the two dice.) If the shooter makes a point on the first roll, then the shooter wins by rolling the same point before rolling a 7 and loses by rolling a 7 before the point. If the shooter's point is 5, what is the probability that the shooter wins the game?*

SOLUTION: This question can be viewed as a conditional probability. If on the next roll the shooter rolls anything but a 5 or 7, the roll has no effect on the game and can be ignored. All that matters is

$$B = \left\{ \begin{array}{l} (1,4),(2,3),(3,2),(4,1), \\ (1,6),(2,5),(3,4),(4,3),(5,2),(6,1) \end{array} \right\}$$

which is the event the shooter rolls a 5 or 7 to end the game. The shooter wins with the event

$$A = \{(1,4),(2,3),(3,2),(4,1)\}$$

so we must calculate $P(A|B)$. Because $A \subset B$, it follows that $A \cap B = A$ and

$$P(A|B) = \frac{\mathrm{n}(A \cap B)}{\mathrm{n}(B)} = \frac{\mathrm{n}(A)}{\mathrm{n}(B)} = \frac{4}{10}.$$

Thus given that the shooter's point is 5, the shooter's probability of winning is 0.4. □

Multiplying both sides of the equation

$$\frac{P(A \cap B)}{P(B)} = P(A|B)$$

by $P(B)$, produces

$$\boxed{P(A \cap B) = P(B)P(A|B).}$$

This turns out to be a very useful formula because in many instances $P(A|B)$ is much easier to compute than $P(A \cap B)$.

Sample Problem 4. *A jar contains three apparently identical coins, but only two of them are fair coins. The probability of tossing tails with the unfair coin is* $1/4$. *If a coin is selected at random and tossed, what is the probability that the unfair coin is selected and the result of the toss is tails? Can we calculate the probability that the result of the toss is tails without knowing which coin is tossed?*

SOLUTION: Let F be the event a fair coin is selected. Hence F' is the event the unfair coin is selected. Let T denote tails. The problem is to compute $P(F' \cap T)$. Clearly $P(F') = 1/3$, and $P(T|F') = 1/4$ follows from the information given about the unfair coin. Therefore, by the above formula

$$\begin{aligned} P(F' \cap T) &= P(F')P(T|F') \\ &= \frac{1}{3} \cdot \frac{1}{4} = \frac{1}{12}. \end{aligned}$$

Although this answers the first question posed, the idea is worth pursuing to answer the more elusive second question. Can we determine the probability of tails without knowing whether or not a fair coin was tossed? Certainly $P(F) = 2/3$ and $P(T|F) = 1/2$. As above

$$\begin{aligned} P(F \cap T) &= P(F)P(T|F) \\ &= \frac{2}{3} \cdot \frac{1}{2} = \frac{1}{3}. \end{aligned}$$

If the outcome of the toss is tails, then either one of the fair coins is tossed and $F \cap T$ occurred or the unfair coin is tossed and $F' \cap T$ occurred. Since $F \cap T$ and $F' \cap T$ are mutually exclusive, $F \cap T$ and $F' \cap T$ form a partition of T and thus

$$P(T) = 1/3 + 1/12 = 5/12$$

by Axiom 3. □

The next section will exploit the method that was just used to find $P(T)$ and answer the posterior question, given that the outcome was tails what is the probability that the unfair coin was used.

Exercises

1. Two fair dice are rolled. What is the probability of doubles given that the sum is at least 10? What is the probability that the sum is 6 given that the number on each die is less than 5? Given that the sum of the numbers on the dice is 6 what is the probability that there is a 2 on at least one of them?

2. Two wheels of chance are spun. The first one is marked with the numbers 1, 2, 3, 4 and the second with 1, 2, 3. On both wheels each number is equally likely. (The sample space is the same as in Exercise 1 from Chapter 1, Section 5.) Compute the following conditional probabilities:

   (a) The sum of the numbers is odd, given a 2 or a 4 on the first wheel,

   (b) at least one number is even, given that the sum is 4,

   (c) the numbers are equal, given that at least one of them is odd.

3. In a sample space with equally likely simple events, A and B are two events. Furthermore, the numbers of elements in certain subsets are $n(A \cup B) = 10, n(B) = 7, n(A \cap B) = 3$. Compute $P(B|A)$.

4. A recent survey shows that 28% of the residents of College Park shop at Safeway, 35% at Giant, and 12% shop at both. A resident is selected at random. Compute the probability of each of the following:

   (a) The resident selected shops at Giant given that he/she shops at Safeway.

   (b) The resident selected does not shop at Safeway given that he/she shops at Giant.

   (c) The resident selected shops at Giant given that he/she shops at least at one of these two stores.

(d) The resident selected shops at Safeway given that he/she does not shop at both.

(e) The resident does not shop at Giant given that he/she does not shop at Safeway.

(f) The resident selected shops at Giant given that he/she shops at exactly one of these two stores.

5. Let A and B be events in a sample space with $P(A) = 0.1, P(B) = 0.2$, and $P(A|B) = 0$. Find $P(A \cup B)$.

6. An urn contains 14 rubies and 10 diamonds; a second urn contains 10 rubies and 2 diamonds. A fair die is rolled. If the outcome of the roll of the die is 1, you may select a gem at random from the first urn. Otherwise, you select a gem at random from the second urn. What is the probability of selecting a diamond? What is the probability of rolling a 1 and selecting a ruby?

7. An urn contains 7 red balls and 4 blue balls. A fair coin is tossed. If it comes up heads, two balls are selected at random from the urn. If it comes up tails, 3 balls are selected at random from the urn. What is the probability of selecting exactly one blue ball?

8. Let C and D be events in a sample space S with $P(C) = 1/3, P(D) = 5/12$, and $P(D|C) = 3/4$. Compute $P(C \cap D)$ and $P(C|D')$.

9. Find $P(C \cap D)$ and $P(C|D)$ if C and D are events in a sample space with $P(C) = 1/2$, $P(D) = 3/8$ and $P(D|C) = 1/4$.

10. A state examining board has the following data on people passing and failing the exam on their first and second attempts of two allowed tries:

	PASS	FAIL
FIRST TRY	15%	25%
SECOND TRY	20%	40%

(a) Given that a random examinee is taking the exam for his/her first time, what is the probability that he/she passes the exam?

(b) Given that a random examinee is taking the exam for his/her second time, what is the probability that he/she fails the exam?

(c) What is the probability that an examinee was taking the exam for his/her first time given that he/she passed the exam?

11. A small college has the following data about the gender and area of major of its students:

	ARTS	SCIENCE
FEMALE	200	150
MALE	250	125

(a) Given that a random student is majoring in a science, what is the probability that the student is female?

(b) Given that a random student is male, what is the probability that the student is not majoring in a science?

3.2 Bayes Theorem

One of the basic rules of probability, Axiom 3, is that the probability of an event can be calculated by partitioning the event into several pieces, calculating the probability of each piece, and then adding. The key idea in this section is to combine Axiom 3 with conditional probabilities, specifically to use partitions of the whole sample for which we can easily compute conditional probabilities. Tree diagrams will also reappear in this section as an effective means of keeping track of the probabilities of the events in the partition and the various conditional probabilities.

Let A be an event in a sample space S and let $B_1, B_2, \ldots, B_n$ be a partition of S. Since each point in A lies in exactly one B_i,

$$A \cap B_1, A \cap B_2, \ldots, A \cap B_n$$

is a partition of A and hence

$$P(A) = P(A \cap B_1) + P(A \cap B_2) + \ldots + P(A \cap B_n).$$

Because for each i

$$P(A \cap B_i) = P(B_i)P(A|B_i)$$

the above can be written

$$P(A) = P(B_1)P(A|B_1) + P(B_2)P(A|B_2) + \ldots + P(B_n)P(A|B_n).$$

The meaning of this equation is that we can calculate the probability of the event A if we know the probability of each B_i in a partition S and the conditional probability of A given each B_i.

Figure 1 shows how a partition B_1, B_2, B_3 of S made with straight lines decomposes both the circular region A and its complement A' into 3 pieces. In this situation the above formula is

$$P(A) = P(B_1)P(A|B_1) + P(B_2)P(A|B_2) + P(B_3)P(A|B_3).$$

Of course, the same reasoning applies to the complement A' of A and for A' the above formula becomes

$$P(A') = P(B_1)P(A'|B_1) + P(B_2)P(A'|B_2) + P(B_3)P(A'|B_3).$$

Although we will work primarily with partitions of S into two or three pieces, the principles and techniques used will work for a partition of S into any finite number of pieces.

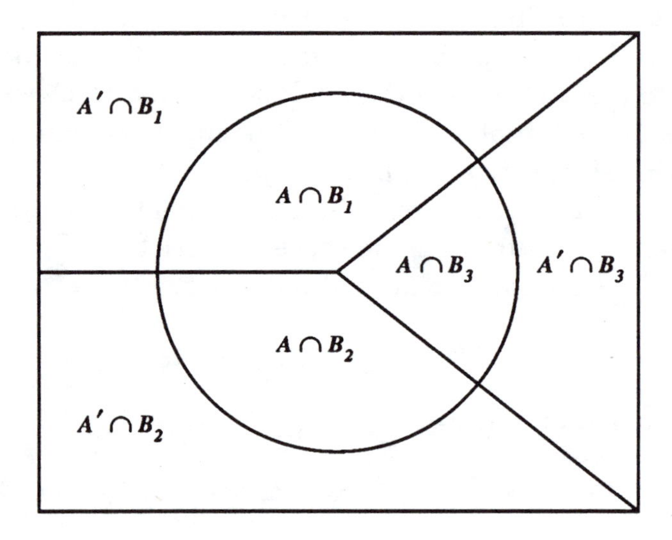

Figure 1

The two formulas

$$P(A) = P(B_1)P(A|B_1) + P(B_2)P(A|B_2) + P(B_3)P(A|B_3)$$

and

$$P(A') = P(B_1)P(A'|B_1) + P(B_2)P(A'|B_2) + P(B_3)P(A'|B_3)$$

contain a lot of information. A convenient way to keep track of all the information in these formulas is to write the information on the branches of a tree diagram as shown in Figure 2.

In a specific problem the actual numerical values of the probabilities would be written on the tree diagram according to the scheme in Figure 2. Each path from the vertex on the extreme left to either A or A' on the right is called a branch. The tree diagram in Figure 2 contains six branches.

The phrase "multiplying along a branch" means simply taking the product of the probabilities written on that branch. The result of multiplying along a branch is always the probability of an intersection of events. For example, by multiplying along the top branch in Figure 2, we get

$$P(B_1)P(A|B_1) = P(A \cap B_1).$$

If we multiply along all the branches ending in A and sum the results, we get $P(A)$. In fact, this process is equivalent to using the formula for $P(A)$ in the beginning of this paragraph. Similarly, the probability of A' can be computed by multiplying along all the branches ending in A' and adding the results.

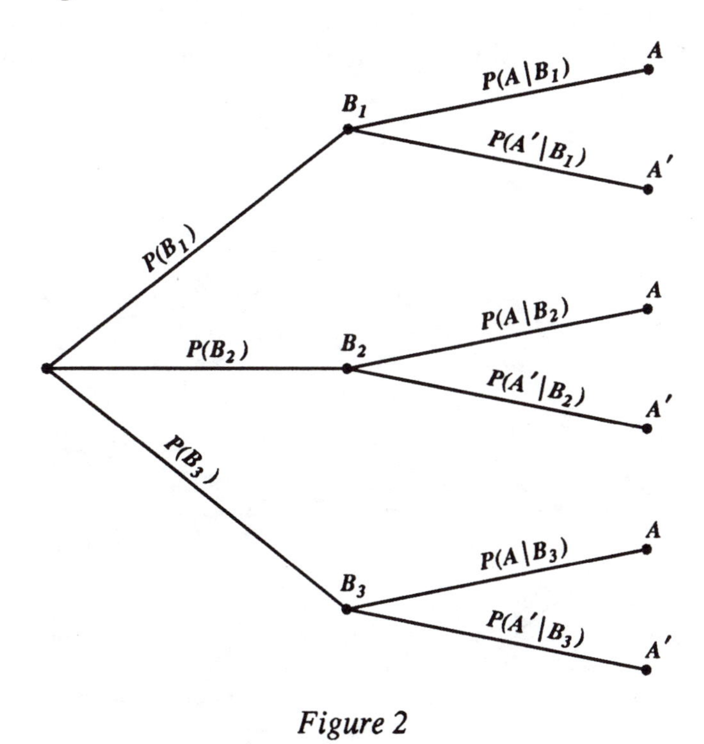

Figure 2

Sample Problem 1. *Suppose a number is selected at random from the numbers 1, 2, and 3 and then a fair coin is tossed that many times.*

(a) *What is the probability of heads on at least one toss?*

(b) *What is the probability that the coin is tossed three times and each time the result is tails?*

(c) *What is the probability of at least one heads with at most two tosses of the coin?*

(d) *Given that heads occurred at least once, what is the probability that the coin was tossed twice?*

SOLUTION: These questions can be answered more easily with the aid of a tree diagram. First notice the number of times the coin is tossed provides a natural partition. Let B_1 be the event the coin is tossed one time, B_2 be the event the coin is tossed two times, and B_3 be the event the coin is tossed three times. Because the numbers are selected at random,

$$P(B_1) = P(B_2) = P(B_3) = 1/3.$$

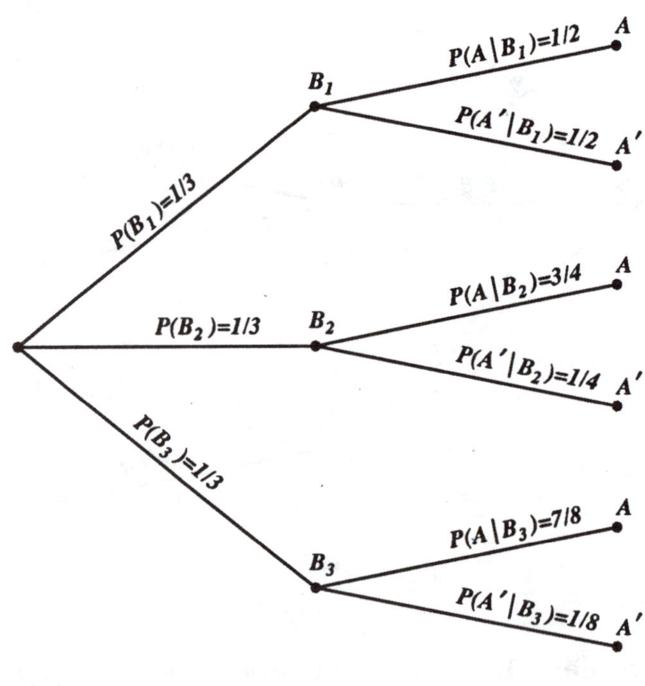

Figure 3

The other event which is important in this problem is getting heads at least once in however many times the coin is tossed. It will be denoted by A. For a fixed number of tosses it is easy to compute the probability of A. For example, tossing a coin 3 times has 8 possible outcomes of which the only one that does not include at least one head is three tails. It follows that $P(A|B_3) = 7/8$ and $P(A'|B_3) = 1/8$. Doing the comparable calculations for the other two cases completes the following table:

$$\begin{array}{ll} P(A|B_1) = 1/2 & P(A'|B_1) = 1/2 \\ P(A|B_2) = 3/4 & P(A'|B_2) = 1/4 \\ P(A|B_3) = 7/8 & P(A'|B_3) = 1/8. \end{array}$$

All these probabilities are written in the appropriate places on the tree diagram in Figure 3.

We are now ready to answer the specific questions.

(a) To compute $P(A)$ multiply along each branch ending in A and add to get

$$P(A) = \frac{1}{3} \cdot \frac{1}{2} + \frac{1}{3} \cdot \frac{3}{4} + \frac{1}{3} \cdot \frac{7}{8} = \frac{17}{24}$$

(b) This is a matter of computing $P(A' \cap B_3)$ which is obtained by multiplying along the bottom branch. Thus

$$P(A' \cap B_3) = \frac{1}{3} \cdot \frac{1}{8} = \frac{1}{24}.$$

(c) Getting heads at least once with at most 2 tosses of the coin is, in our notation, the same as $(A \cap B_1) \cup (A \cap B_2)$. To compute this probability multiply along the branches through B_1 and B_2 which end in A and then add. Consequently,

$$P[(A \cap B_1) \cup (A \cap B_2)] = \frac{1}{3} \cdot \frac{1}{2} + \frac{1}{3} \cdot \frac{3}{4} = \frac{5}{12}$$

(d) In this part the problem is to compute the probability that the coin was tossed twice given that heads occurred at least once, that is, $P(B_2|A)$. From the definition of conditional probability

$$P(B_2|A) = \frac{P(B_2 \cap A)}{P(A)}$$

Note that the denominator was calculated in part (a) and the numerator can be found by multiplying along the correct branch. Thus

$$\begin{aligned} P(B_2|A) &= \frac{P(B_2)P(A|B_2)}{P(A)} \\ &= \frac{\frac{1}{3}\cdot\frac{3}{4}}{\frac{17}{24}} = \frac{1}{4}\cdot\frac{24}{17} = \frac{6}{17} \end{aligned}$$

which is called a posterior probability because the condition is the final outcome.

This completes the solution. □

The solution to part (d) above is just a special case of a general principle. Let A and B be events in a sample space S. We can speak of the conditional probability of A given B and vice versa, and generally speaking, they are not equal. Observe that in the formulas

$$P(A|B) = \frac{P(A \cap B)}{P(B)}$$

and

$$P(B|A) = \frac{P(A \cap B)}{P(A)}$$

$P(A \cap B)$ appears in both of them. The result of multiplying the first by $P(B)$ and the second by $P(A)$ is

$$P(B)P(A|B) = P(A \cap B)$$

$$P(A)P(B|A) = P(A \cap B).$$

Therefore,

$$\boxed{P(B)P(A|B) = P(A)P(B|A).}$$

which is the heart of Bayes' Theorem. The principle that was alluded to at the beginning of the paragraph can now be stated; any time three of the four terms in this equation are known, the fourth can be determined.

Sample Problem 2. *Let E and F be events in a sample space S with $P(E) = 0.6, P(F) = 0.4,$ and $P(E|F) = 0.3$. Determine $P(F|E)$.*

SOLUTION: The above formula written with E's and F's is

$$P(E)P(F|E) = P(F)P(E|F).$$

When the given data are substituted, this becomes

$$(0.6)P(F|E) = (0.4)(0.3).$$

Thus

$$P(F|E) = \frac{(0.4)(0.3)}{0.6} = \frac{0.12}{0.6} = 0.2$$

and we have answered the question by solving for the one unknown term. □

Continuing in this vein, suppose B_1, B_2, B_3 is a partition of S and in the previous framed formula replace B with B_i to get

$$P(B_i)P(A|B_i) = P(A)P(B_i|A).$$

Now divide both sides by $P(A)$.

$$\frac{P(B_i)P(A|B_i)}{P(A)} = P(B_i|A).$$

Next from the first page of this section use the formula

$$P(A) = P(B_1)P(A|B_1) + P(B_2)P(A|B_2) + P(B_3)P(A|B_3).$$

to substitute for $P(A)$. This gives us

$$\boxed{P(B_i|A) = \frac{P(B_i)P(A|B_i)}{P(B_1)P(A|B_1) + P(B_2)P(A|B_2) + P(B_3)P(A|B_3)}}$$

which is *Bayes' Theorem.* It is usually preferable to use Bayes' Theorem in two steps - first calculate the denominator with a tree diagram and then divide this answer into the result of multiplying along the correct branch for the numerator. This process is illustrated in the next Sample Problem.

Sample Problem 3. *Three urns contain red and green balls. The first urn contains 2 red and 10 green balls, the second urn contains 5 red and 15 green balls, and the third urn contains 12 red and 6 green balls.*

A fair die is rolled and the outcome is used to select an urn. When a 1 comes up on the die, the first urn is selected; when a 2 or 3 comes up on the die, the second urn is selected; when a 4, 5, or 6 comes up on the die, the third urn is selected. After selecting the urn, a ball is randomly selected from it. Given that a green ball was selected, what is the probability that it came from the first urn?

SOLUTION: The first step is to carefully sort and label the information. Let U_1, U_2, U_3 be the events that the ball is selected from the first, second, and third urns respectively. This is the natural partition to use for the problem. Let G be the event a green ball is selected. The problem asks us to determine $P(U_1|G)$, which is the same type of problem that was solved in part (d) of Sample Problem 1. It is important to recognize that an essential part of the solution is the determination of $P(G)$.

One way to attack this problem is to start with

$$P(G)P(U_1|G) = P(U_1)P(G|U_1).$$

Obviously $P(U_1) = 1/6$ which is the probability of rolling a 1 with a fair die. Furthermore, $P(G|U_1) = 10/12$ because the first urn contains 12 balls of which 10 are green. Thus two of the four probabilities in this formula are immediately available from the given data. Since

$$P(U_1|G) = \frac{P(U_1)P(G|U_1)}{P(G)} = \frac{\frac{1}{6} \cdot \frac{10}{12}}{P(G)},$$

the solution can be completed by using a tree diagram to compute

$P(G)$. From the tree diagram (Figure 4) it follows that

$$\begin{aligned} P(G) &= \frac{1}{6}\cdot\frac{10}{12}+\frac{1}{3}\cdot\frac{15}{20}+\frac{1}{2}\cdot\frac{6}{18} \\ &= \frac{5}{36}+\frac{1}{4}+\frac{1}{6}=\frac{20}{36}=\frac{5}{9}. \end{aligned}$$

Thus

$$P(U_1|G)=\frac{\frac{1}{6}\cdot\frac{10}{12}}{\frac{5}{9}}=\frac{5}{36}\cdot\frac{9}{5}=\frac{1}{4}.$$

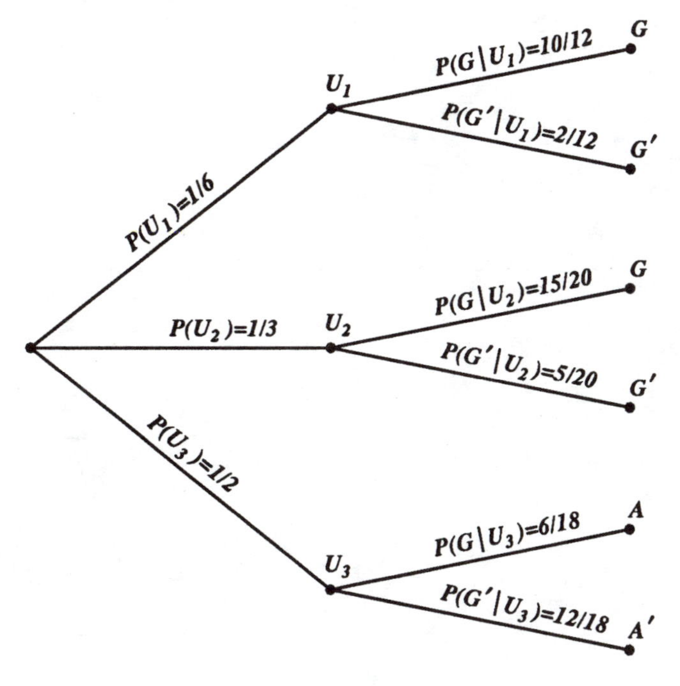

Figure 4

Alternatively we could have substituted directly into Bayes' Theorem as follows:

$$\begin{aligned} P(U_1|G) &= \frac{P(U_1)P(G|U_1)}{P(U_1)P(G|U_1)+P(U_2)P(G|U_2)+P(U_3)P(G|U_3)} \\ &= \frac{\frac{1}{6}\cdot\frac{10}{12}}{\frac{1}{6}\cdot\frac{10}{12}+\frac{1}{3}\cdot\frac{15}{20}+\frac{1}{2}\cdot\frac{6}{18}} \end{aligned}$$

$$= \frac{\frac{5}{36}}{\frac{5}{36} + \frac{1}{4} + \frac{1}{6}} = \frac{\frac{5}{36}}{\frac{20}{36}} = \frac{1}{4}$$

and obtained the same answer. □

Exercises

1. Three urns contain red and green colored marbles as follows: urn 1 has 3 green and 5 red, urn 2 has 9 green and 7 red, urn 3 has 15 green and 17 red. If an urn is selected at random and a marble is then selected at random from this urn, what is the probability that a green marble is selected? What is the probability that the marble selected comes from urn 1 and is red? What is the probability that the marble selected does not come from the first urn and is red?

2. There are three drawers containing socks. The first contains 5 red and 3 yellow socks. The second contains 1 red and 3 yellow socks. The third contains 3 red and 5 yellow socks. A drawer is chosen at random and a sock is selected. What is the probability that the sock is yellow?

3. A box contains 4 five dollar bills and 6 ten dollar bills. A bill is randomly selected from the box and laid aside without anyone seeing it. Then a second bill is selected from the box. What is the probability that the second bill selected is a five dollar bill?

4. An urn contains 3 black and 2 white balls. A ball is randomly selected from the urn and laid aside without anyone seeing it. A second ball is then randomly selected from the urn. What is the probability that the second ball selected is black?

5. Let E and F be events in a sample space S with $P(E) = 0.96$, $P(F) = 0.6$, and $P(E|F) = 0.4$. Find $P(F|E)$

6. Determine $P(A|B)$ when $P(A) = 0.5$, $P(B) = 0.9$, and $P(B|A) = 0.72$.

7. Find $P(A)$ if A and B are events in a sample space with $P(A|B) = 0.8$, $P(B|A) = 0.6$, and $P(B) = 0.3$.

8. Use Figure 4 to compute the probability that the ball selected in Sample Problem 3 came from the third urn given that it is red.

9. Two jars contain cherry and orange candies. The first jar contains 5 cherry and 1 orange candies and the second jar contains 4 cherry and 8 orange candies. A jar is selected at random and a candy is drawn at random from it. What is the probability that an orange candy was selected? Given that a cherry candy was selected, what is the probability that it came from the first jar?

10. Three bowls numbered 0, 1, and 2 contain red and blue chips. The bowl numbered 0 contains 12 blue and 4 red chips; the bowl numbered 1 contains 14 blue and 2 red chips; and the bowl numbered 2 contains 4 blue and 28 red chips. A fair coin is tossed twice and a chip is drawn at random from the bowl whose number is the same as the number of heads obtained from tossing the coin.

    (a) What is the probability of selecting a blue chip?

    (b) What is the probability of getting at most one head and selecting a red chip?

    (c) Given at least one head on the coin toss, what is the probability of selecting a red chip?

    (d) What is the probability that the chip selected came from the bowl numbered 1, given that it is blue?

11. The probability that it will be sunny this weekend is $4/5$. Given that it is sunny the probability you will go to the beach is $1/4$. The probability that you go to the beach is $3/10$. Given that you went to the beach what is the probability that it was sunny?

12. An urn contains 3 green and 5 yellow marbles, and a second urn contains 7 green and 2 yellow marbles. A marble is randomly selected from the first urn and transferred to the second urn without noting its color. Then a marble is randomly selected from the second urn. What is the probability that the marble selected from the second urn is yellow?

13. A bag contains 3 half dollars and 2 silver dollars. Two coins are randomly selected at once from this bag and transferred to

a second bag which already contains 4 half dollars and 2 silver dollars. A coin is then drawn from the second bag. Assuming you do not know the values of the coins transferred, compute the probability that the coin selected from the second bag is a silver dollar. Compute the probability that 2 half dollars were transferred given that a silver dollar was selected from the second bag.

14. On a quiz show, three boxes each contain 25 envelopes. In the first box, 24 of the envelopes each contain 100 dollars; in the second box, 15 contain 100 dollars; and in the third 12 contain 100 dollars. The rest of the envelopes are empty. A contestant rolls a fair die and selects an envelope at random from the first box if he rolls a 1, from the second box if he rolls a 2 or 3, and from the third box if he rolls a 4, 5, or 6. What is the probability that the contestant selected an envelope from the first box, given that he won 100 dollars? Given that the contestant did not win 100 dollars, what is the probability that he selected an envelope from the third box?

15. The probability that a student is a freshman is 1/4. Given that a student is a freshman, the probability that he/she takes Math. 111 is 1/8. The probability that a student takes Math. 111 is 1/30. Given that a student takes Math. 111, what is the probability that he/she is a freshman?

16. A bottle contains 12 pills, identical in appearance, but only 4 are medicine. The other 8 are sugar placebos with no medicinal value at all. Jan selects a pill at random and swallows it. If the pill Jan takes is medicine, the probability of relief from the headache is .9. If the placebo is taken, the probability of relief is .2. What is the probability of relief from the headache? What is the conditional probability that the pill taken was medicine, given that the headache is relieved?

17. In oil exploration past experience has shown that the probability of a successful strike is .1. Geologists have observed that if there is oil, the probability is .6 that porous rock is present. However,

records indicate that when no oil is below, the probability is .2 that such rock is present.

a) What is the probability that porous rock will be found in oil exploration drillings?

b) Given that porous rock is found what is the probability of striking oil?

18. A jury consisting of 4 men and 7 women is assigned a twelfth member by randomly selecting a name from a pool of prospective jurors consisting of 3 men and 2 women. If the foreman is then chosen at random, what is the probability that the foreman is a woman?

19. An urn contains 2 pearls and nothing else. A fair die is rolled. If the outcome of the roll of the die is a one or a two, 2 white beads are added to the urn; if the outcome is a five or a six, 6 white beads are added. Then an object is randomly selected from the urn. What is the probability of selecting a pearl? Given that a pearl was selected what is the probability that only 2 white beads were added to the urn?

20. A commercial fitness program claims that 60% of the male population between the ages of 35 and 55 gets too little exercise and that those who get too little exercise are twice as likely to have a heart attack as the others. They also claim the probability that a male between 35 and 55 who gets enough exercise will have a heart attack is .001. Assuming their figures are correct what is the probability that a male heart attack victim between 35 and 55 did not get enough exercise?

21. A blue cookie jar contains 3 chocolate chip cookies, 4 almond cookies and 1 oatmeal cookie. A green cookie jar contains 7 chocolate chip cookies, 5 almond cookies and 4 oatmeal cookies. One of the jars is selected at random and a cookie drawn from it.

a) What's the probability that the cookie drawn is an almond cookie?

b) What is the probability that an oatmeal cookie is selected from the green cookie jar?

c) Given that the cookie selected is a chocolate chip cookie, what is the probability that it came from the blue jar?

22. Suppose a diagnostic test for cancer is 98% accurate, and it is being used in a high risk population for which the probability that a random individual has cancer is 0.5%. What is the probability that a random person from this population will test positive? Given that an individual tested positive, what is the probability that he or she has cancer?

3.3 Independence

Sometimes knowing that an event B has occurred has absolutely no effect on the probability of another event A; that is, the occurrence of B neither increases nor decreases the probability of A. The most familiar example of this is the probability that the second toss of a fair coin is heads given that the first toss was heads. Using the sample space

$$\begin{aligned} S &= \{HH, HT, TH, TT\}, \\ A &= \{HH, TH\} = \{\text{heads on second toss }\} \\ B &= \{HH, HT\} = \{\text{ heads on first toss }\} \\ A \cap B &= \{HH\} = \{\text{ heads on both tosses }\} \end{aligned}$$

and

$$P(A) = \frac{1}{2} = P(B).$$

Then given that B has occurred, the probability of A is

$$P(A|B) = \frac{\mathbf{n}(A \cap B)}{\mathbf{n}(B)} = \frac{1}{2} = P(A).$$

Even if the coin is tossed 99 times and comes up heads every single time, the probability of heads on the 100th toss is still 1/2. In fact, the probability of heads on the 100th toss is 1/2 no matter what happened on the first 99 tosses; the outcome of the 100th toss is independent of all previous tosses.

The phenomenon we just described is usually called independence and the formal definition follows. Two events A and B in a sample space S are *independent* if

$$P(A|B) = P(A).$$

Although the idea behind independence is clearly evident in this formula, it is not the easiest formula to use. Because

$$P(A|B) = \frac{P(A \cap B)}{P(B)},$$

the above can be written

$$\frac{P(A \cap B)}{P(B)} = P(A).$$

After multiplying both sides by $P(B)$, we see that A and B are independent if

$$\boxed{P(A \cap B) = P(A)P(B).}$$

Usually the easiest way to determine whether or not two events A and B are independent is to compute both $P(A \cap B)$ and $P(A)P(B)$ and then compare the results. If they are equal, the events are independent. Otherwise they are said to be *dependent events.*

Sample Problem 1. *Consider the following events when two fair dice - a red and a green one - are rolled:*

(a) doubles,

(b) 3 on the green die,

(c) 5 points.

Which pair(s) of these three events are independent?

SOLUTION: Using the usual 36 element sample space for two dice and using A, B, C to denote the events described in parts (a), (b), and (c), we have

$$S = \left\{ \begin{array}{l} (1,1),(1,2),(1,3),(1,4),(1,5),(1,6), \\ (2,1),(2,2),(2,3),(2,4),(2,5),(2,6), \\ (3,1),(3,2),(3,3),(3,4),(3,5),(3,6), \\ (4,1),(4,2),(4,3),(4,4),(4,5),(4,6), \\ (5,1),(5,2),(5,3),(5,4),(5,5),(5,6), \\ (6,1),(6,2),(6,3),(6,4),(6,5),(6,6) \end{array} \right\}$$

$$\begin{aligned} A &= \{(1,1),(2,2),(3,3),(4,4),(5,5),(6,6)\}, \\ B &= \{(1,3),(2,3),(3,3),(4,3),(5,3),(6,3)\}, \\ C &= \{(1,4),(2,3),(3,2),(4,1)\}, \end{aligned}$$

where as before the first number comes from the red die and the second comes from the green die. It follows that

$$P(A) = P(B) = 1/6,$$

and

$$P(C) = 1/9.$$

There are three pairs of events to consider - A and B, B and C, A and C. Because $A \cap B = \{(3,3)\}$, $P(A \cap B) = 1/36$ which is the same as $P(A)P(B) = (1/6)(1/6)$. Thus A and B are independent events. Next consider the pair B and C. Now $B \cap C = \{(2,3)\}$ and $P(B \cap C) = 1/36$, but $P(B)P(C) = (1/6)(1/9) = 1/54$. So B and C are not independent and hence dependent events. The last pair is A and C. Because $A \cap C = \emptyset$,

$$P(A \cap C) = 0 \neq (1/6)(1/4) = P(A)P(C)$$

and they are also dependent events. □

The distinction between disjoint and independent events merits special attention. These two concepts which are often confused by students are very different. In fact independent events with positive probabilities are never disjoint. To understand why this is true consider independent events E and F with positive probabilities. Then

$$P(E \cap F) = P(E)P(F) > 0$$

Suppose E and F were also disjoint, then $E \cap F = \emptyset$ would imply that

$$P(E \cap F) = P(\emptyset) = 0$$

which is incompatible with $P(E \cap F) > 0$.

Continuing to work with independent events, we can ask what does the independence of E and F imply about $P(E \cap F')$? Using several of the formulas at the end of Section 1.4 and the formula $P(E \cap F) = P(E)P(F)$ we see that

$$\begin{aligned} P(E \cap F') &= P(E) - P(E \cap F) \\ &= P(E) - P(E)P(F) \\ &= P(E)[1 - P(F)] \\ &= P(E)P(F'), \end{aligned}$$

and E and F' are also independent events.

Likewise a similar calculation shows that E' and F' are independent events when E and F are independent. Starting with de Morgan's formula and using a little more algebra, we have

$$\begin{aligned} P(E' \cap F') &= 1 - P(E \cup F) \\ &= 1 - [P(E) + P(F) - P(E \cap F)] \\ &= 1 - P(E) - P(F) + P(E)P(F) \\ &= [1 - P(E)][1 - P(F)] \\ &= P(E')P(F'). \end{aligned}$$

Therefore, if E and F are independent events, then E and F' are independent, E' and F are independent, and E' and F' are independent events. This fact makes it very easy to fill in a Venn diagram for two independent events.

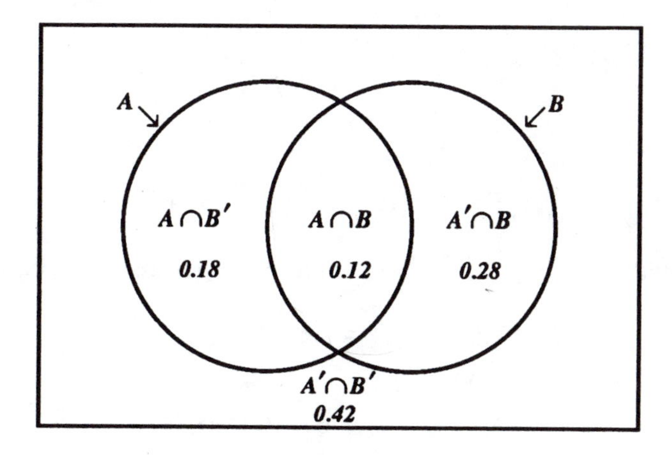

Figure 1

EXAMPLE 1. If A and B are two independent events in a sample space S with $P(A) = 0.3$ and $P(B) = 0.4$, then

$$\begin{aligned} P(A \cap B) &= (0.3)(0.4) = 0.12 \\ P(A \cap B') &= (0.3)(0.6) = 0.18 \\ P(A' \cap B) &= (0.7)(0.4) = 0.28 \end{aligned}$$

$$P(A' \cap B') = (0.7)(0.6) = 0.42$$

and Figure 1 shows the corresponding Venn diagram.

The correct definition of independence for three or more events requires not only that all pairs of the events be independent, but also that each event be independent from intersections of the other events. The simplest way to state this is that the probability of any intersection equals the product of the probabilities of the events intersected. For example, if A and $B \cap C$ are independent events and if B and C are independent events, then

$$\begin{aligned} P(A) &= P(A|B \cap C) \\ &= \frac{P(A \cap B \cap C)}{P(B \cap C)} \\ &= \frac{P(A \cap B \cap C)}{P(B)P(C)} \end{aligned}$$

which implies that

$$P(A)P(B)P(C) = P(A \cap B \cap C).$$

For three events A, B and C this means independence is equivalent to

$$\begin{aligned} P(A \cap B) &= P(A)P(B) \\ P(A \cap C) &= P(A)P(C) \\ P(B \cap C) &= P(B)P(C) \end{aligned}$$

and

$$P(A \cap B \cap C) = P(A)P(B)P(C)$$

Unfortunately the last equation does not imply the first three or vice versa which partially explains why the definition of independence for two events is so much simpler than for three or more events.

The events $A_1, A_2, \ldots, A_k$ in a sample space S are independent provided that for any subset $\{i, \ldots, j\}$ of $\{1, 2, 3, \ldots, k\}$

$$P(A_i \cap \ldots \cap A_j) = P(A_i) \cdot \ldots \cdot P(A_j)$$

Earlier in this section we showed that the independence of two events A and B implied the independence of the following pairs of events: A and B', A' and B, A' and B'. The analogue of this fact for independent events $A_1, \ldots, A_k$ is that any number of these events can be replaced by their complements without destroying the independence. The practical consequence of this is that when you are working with independent events, probabilities of intersections can be computed by multiplying the individual probabilities.

EXAMPLE 2. If A_1, A_2, A_3, and A_4 are independent events, then

$$P(A_1 \cap A_2' \cap A_3' \cap A_4) = P(A_1)P(A_2')P(A_3')P(A_4).$$

More often than not independence is assumed, not checked. Anytime there is good reason to believe that the outcomes of several events are in no way influenced by each other, it is reasonable to assume that the events are mutually independent. This has the decided advantage of determining the probabilities of all intersections from the probabilities of the main events. Once the probabilities of intersections are known, then probabilities of unions are readily computed. The point is that independence makes many problems more tractable and the initial analysis of a situation as to whether or not there is reason to assume events are independent is crucial. In practice the assumption of independence is often a matter of human judgement.

Sample Problem 2. *Three different radar systems are deployed at separate sites in a region. They have probabilities of* 0.9, 0.8, *and* 0.7 *of detecting an enemy aircraft in this region.*

(a) What is the probability that only the first system would detect an enemy aircraft?

(b) What is the probability that none of them would detect an enemy aircraft?

(c) What is the probability that exactly two of the systems would detect an enemy aircraft?

SOLUTION: Denote the events that the first, second, and third radar systems detect an enemy aircraft by A, B, and C respectively, so $P(A) = 0.9$, $P(B) = 0.8$ and $P(C) = 0.7$. The wording of the problem definitely suggests that these three radar systems are independent of each other; hence it is reasonable to assume that A, B and C are independent events.

(a) The probability that only the first system detects an enemy aircraft is the same as $P(A \cap B' \cap C')$. Because the independence of A, B and C implies the independence of A, B' and C',

$$\begin{aligned} P(A \cap B' \cap C') &= P(A)P(B')P(C') \\ &= (0.9)(0.2)(0.3) = 0.054. \end{aligned}$$

(b) Here we must compute $P(A' \cap B' \cap C')$ and as in part (a)

$$\begin{aligned} P(A' \cap B' \cap C') &= P(A')P(B')P(C') \\ &= (0.1)(0.2)(0.3) = 0.006. \end{aligned}$$

(c) Since the question does not specify which two of the systems will detect the enemy aircraft, the three possibilities - $A \cap B \cap C'$, $A \cap B' \cap C$ and $A' \cap B \cap C$ - must be included in the answer. Because these three events are mutually exclusive, the answer is

$$(0.9)(0.8)(0.3) + (0.9)(0.2)(0.7) + (0.1)(0.8)(0.7) = 0.398.$$

Although this finishes the solution of the problem as posed, it is straight forward to complete a Venn diagram for these three events (Figure 2) and to answer other questions such as determining the probability that at most one radar station detects an enemy aircraft. □

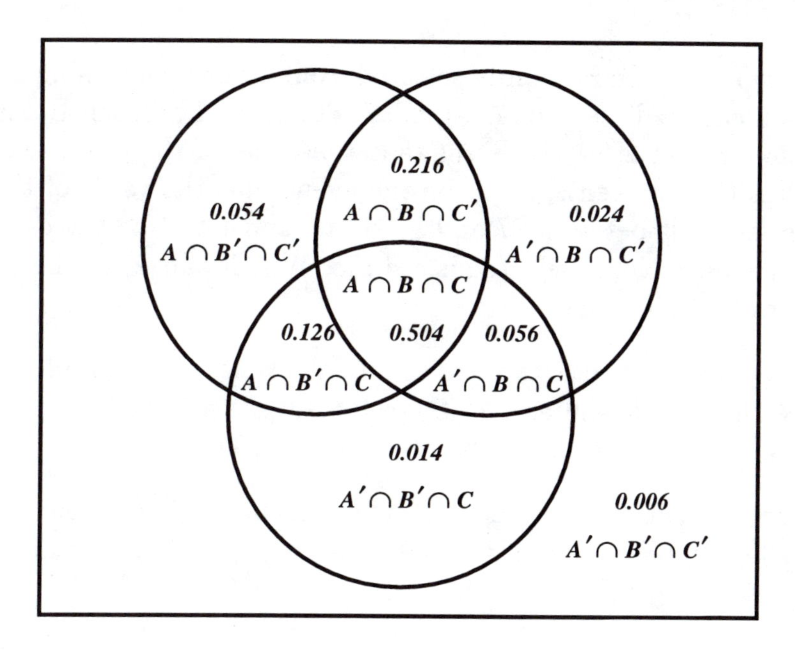
0.216
$A \cap B \cap C'$
0.054
$A \cap B' \cap C'$
0.024
$A' \cap B \cap C'$
$A \cap B \cap C$
0.126
0.504
0.056
$A \cap B' \cap C$
$A' \cap B \cap C$
0.014
$A' \cap B' \cap C$
0.006
$A' \cap B' \cap C'$

Figure 2

Exercises

1. Two coins are randomly selected from an urn containing a penny, a nickel, a dime, a quarter, a half dollar, and a Susan B. Anthony dollar. Let A be the event that the value of the coins selected is less than 35 cents, let B be the event that the value of the coins selected is between 27 and 52 cents, and let C be the event that one of the coins selected is a dime. Which pair(s) of these events are independent?

2. Suppose C and D are independent events in a sample space S with $P(C) = 0.6$ and $P(D) = 0.8$. Make a Venn diagram for C and D showing the probabilities of each cell.

3. Let A and B be independent events in a sample space S with $P(A) = 0.3$ and $P(B) = 0.9$. Compute the following probabilities:

   (a) $P(A \cap B)$

   (b) $P(A \cup B)$

   (c) $P(A' \cap B)$.

4. A fair coin is tossed 3 times. Let D be the event of heads on the second toss, let E be the event of heads on at most one toss, and let F be the event of tails on the first and last tosses. Which pair(s) of these events are independent?

5. The probability your car will start in the morning is .8. The probability your roommate's car will start is .6. Compute the probabilities of the following events in a particular morning:

   (a) exactly one of the two cars starts,

   (b) both cars start,

   (c) neither car starts.

6. An urn contains four balls colored red, yellow, blue, and green. A ball is randomly selected from this urn. Let A be the event that the red or green ball is selected, let B be the event that the

yellow or green ball is selected and let C be the event that the blue or green ball is selected.

(a) Check that A and B, A and C, and B and C are pairs of independent events.

(b) Calculate $P(A \cap B \cap C)$ and $P(A)P(B)P(C)$.

(c) Are the three events A, B, and C independent?

(d) Calculate $P(A|B \cap C)$. Does it equal $P(A)$?

(e) Are A and $B \cap C$ independent?

7. For the situation described in Sample Problem 2 of the text, use Figure 2 to answer the following questions:

   (a) What is the probability that exactly one of these radar systems would detect an enemy aircraft?

   (b) What is the probability that at most one of them would detect an enemy aircraft?

8. The probabilities that a certain student will be assigned a term paper to be written over spring break are 0.7 in his economics course and 0.8 in his history course.

   (a) What is the probability that the student will have to write a term paper in both of these courses during the spring break?

   (b) What is the probability that he will have to write a paper in exactly one of these courses during spring break?

   (c) What is the probability that he will not have to write a paper in either course during spring break?

9. The probability that the chairman of a charity fund raising event will make an error counting the receipts is 1/3, so he asks two friends to each recount the money. The probability that the first friend makes an error is 1/4 and the probability that the second friend makes an error is 1/5.

   (a) What is the probability that all three of them make an error?

(b) What is the probability that no one makes an error?

(c) What is the probability that the chairman makes an error and the two friends do not make an error?

10. A number is selected at random from $S = \{1, 2, 3, 4, 5, 6, 7, 8\}$. Let $A = \{1, 2, 3, 4\}$, $B = \{1, 5, 6, 7\}$, and $C = \{1, 2, 5, 8\}$.

    (a) Which pair(s) of these events are independent?

    (b) Verify that $P(A \cap B \cap C) = P(A)P(B)P(C)$.

    (b) Are A, B, and C independent events?

11. A sample space contains the events A and B satisfying $P(A) = 0.4$, $P(B) = 0.6$, and $P(A \cup B) = 0.76$.

    (a) Find $P(A \cap B)$.

    (b) Determine whether or not A and B are independent.

    (c) Compute $P(A|A \cup B)$.

12. Twenty balls numbered $1, 2, \ldots, 20$ are placed in a bowl and one ball is drawn at random. Let A be the event an even number is drawn, B be the event a number which is a multiple of 5 is drawn, and C be the event that a 3, 4, 5, 6, or 7 is drawn. Which pair(s) of these events are independent?

13. If A, B and C are three independent events with $P(A) = 1/2$, $P(B) = P(C)$, and $P(A \cap B \cap C) = 1/18$ find $P(C)$.

14. An urn contains 5 balls numbered 1 to 5 inclusive. A ball is drawn at random, its number written down, and it is put back in the urn. This is repeated 8 times. What is the probability of getting a number greater than 3 exactly 6 times?

15. Sport City, U.S.A. has professional baseball, football, and ice hockey teams. The probabilities that they will win their respective championships (World Series, Super Bowl, and Stanley Cup) are 0.5, 0.2, and 0.5. Fill in a Venn diagram for this situation and answer the following questions:

(a) What is the probability that none of their teams will win a championship?

(b) What is the probability that exactly one of their teams will win a championship?

(c) What is the probability that a least 2 will win championships?

16. There are three identical smoke detectors on the ceiling in a hotel corridor. Each one will fail to work in the event of a fire with probability 1/4. What is the probability that at least one will work in the event of fire? What is the probability that all of them will work in the event of fire? What is the probability that exactly one will work in the event of fire?

17. An army hospital maintains three completely separate electric generators to run its operating room. At any given time the probabilities that they individually fail to generate sufficient power to run the operating room are 0.1, 0.2, and 0.4 respectively.

    (a) What is the probability that at any given time at least one of the generators would generate sufficient power to run the operating room?

    (b) What is the probability that the first two fail to generate sufficient power but the third one does not fail?

    (c) What is the probability that at least two of the generators would fail to generate sufficient power?

3.4 Independent Bernoulli Trials

A *Bernoulli trial* is an experiment with exactly two unpredictable outcomes usually called success and failure. The probability of success will always be denoted by p and the probability of failure by $q = 1 - p$. Which outcome is called success depends on the viewpoint of the person doing the experiment. Success and failure are interchangeable. Two people can analyze the same problem from opposite points of view, and they will obtain the same numerical answers. So it is important to decide at the beginning of a problem which of the two outcomes will be called success and stick with this initial choice.

EXAMPLE 1. The most familiar Bernoulli trial is tossing a fair coin. Success is either heads or tails as called and $p = q = 1/2$.

A Bernoulli trial is in some sense the most primitive random phenomenon because an unpredictable experiment must have at least two outcomes. Tossing a two headed coin is a predictable experiment! Bernoulli trials are also very prevalent because any event can be viewed as a Bernoulli trial. Let S be the sample space for some random phenomenon and let E be the event of interest in S. To create a Bernoulli trial using E call the occurrence of E success and the occurrence of E' failure. In other words ignore all the events in S except E and E'. Then $p = P(E)$ and $q = 1 - p = 1 - P(E) = P(E')$.

EXAMPLE 2. Let S be the usually sample space of 36 elements for rolling a pair of dice and let

$$E = \{(1,6),(2,5),(3,4),(4,3),(5,2),(6,1),(5,6),(6,5)\}.$$

Here E is the event of rolling a seven or eleven which is important in the game of craps. Rolling a seven or eleven can be thought of as a Bernoulli trial with $p = 8/36 = 2/9$.

The more interesting Bernoulli trials are those that can be repeated as often as we like without changing the probability of success, p. The simplest example is tossing a coin. Whether or not a particular baseball player gets a hit when at bat is not a good example of a repeatable Bernoulli trial because the probability of success will vary from day to day and it can not be easily repeated an arbitrary number of times like a coin toss can.

Whenever an experiment is repeated it is almost always desirable to arrange things so that each time the outcome is independent of all previous outcomes. Bernoulli trials are no exception. Repeatedly tossing a coin, rolling a die, and spinning a wheel are naturally independent with minimal precautions. Cards are shuffled between games to ensure that the player's hands are independent from those of previous games. Pollsters must be careful to make sure that the individuals they contact are randomly selected so their preferences can be assumed to be independent of each other. In these and many other situations appropriate steps can be taken so that it is reasonable to assume that the outcomes of a repeated Bernoulli trial are mutually independent. In this book we will always assume that the outcomes of a repeated Bernoulli trial are mutually independent.

When repeating a Bernoulli trial we want to keep track of the successes and failures on the different tries. Success on the ith try will be denoted by A_i and its complement A_i' will indicate failure on the ith try. Because the standing assumption that the outcomes of a repeated Bernoulli trial are mutually independent, the events $A_1, A_2, \ldots, A_k$ are independent and probabilities of intersections can be calculated by multiplying the respective probabilities together as in:

$$P(A_1 \cap A_2' \cap A_3) = p \cdot q \cdot p = p^2 \cdot q.$$

EXAMPLE 3. A pair of dice is rolled until the sum of the faces is ten. Hence in the usual sample space for a pair of dice success is the event

$$\{(4,6),(5,5),(6,4)\}$$

and $p = 3/36 = 1/12$. The probability that the first success occurs on the fifth try is

$$P(A_1' \cap A_2' \cap A_3' \cap A_4' \cap A_5) = q^4 \cdot p = \left(\frac{11}{12}\right)^4 \cdot \frac{1}{12} = 0.0588$$

This means that almost 6% of the time the first ten would occur on precisely the fifth try.

The above example illustrates a general fact. When a Bernoulli trial is repeated until the first success occurs, there is a formula for the probability that the first success occurs on the kth try. In this case there are $k-1$ failures followed by a success which is the same as the event

$$A'_1 \cap A'_2 \cap \ldots \cap A'_{k-1} \cap A_k$$

Because the events A_i are mutually independent it follows that

$$P(A'_1 \cap A'_2 \cap \ldots \cap A'_{k-1} \cap A_k) = q^{k-1} \cdot p.$$

In the above example $p = 1/12$, $q = 11/12$, $k = 5$, and $k - 1 = 4$.

Sample Problem 1. *In the game of Parcheesi a player must roll doubles with a pair of dice to enter one of four playing pieces on the board. In particular, at the beginning of the game players can do nothing until they roll their first doubles. The following questions are concerned with how many turns it takes for a player to get the first piece on the board.*

(a) *What is the probability that it takes a player ten turns to put his or her first piece on the board?*

(b) *What is the probability of entering the board in less than 4 turns?*

(c) *What is the probability that it takes more than three turns to get started?*

SOLUTION: In all three parts the Bernoulli trial is rolling doubles, i.e. success is the event

$$E = \{(1,1),(2,2),(3,3),(4,4),(5,5),(6,6)\},$$

$p = 6/36 = 1/6$ and $q = 5/6$, and it is repeated until the first success occurs.

(a) The first question is answered by using the above formula with $k = 10$. So the probability of rolling doubles for the first time on the tenth try is

$$\left(\frac{5}{6}\right)^9 \cdot \frac{1}{6} = 0.0323.$$

(b) Taking less than four tries to get the first doubles is the same as rolling doubles for the first time on the first, second, or third try. These events are disjoint, so we can calculate the probability of each one of them with the formula and then add.

$$P(A_1) = \frac{1}{6} = 0.1667$$

$$P(A_1' \cap A_2) = \left(\frac{5}{6}\right)^1 \cdot \frac{1}{6} = 0.1389$$

$$P(A_1' \cap A_2' \cap A_3) = \left(\frac{5}{6}\right)^2 \cdot \frac{1}{6} = 0.1157.$$

The sum of these three numbers is 0.4213. and it is rather likely that one of four players will get started quickly.

(c) The last part is best done by calculating the probability of the complement. The complement of more than 3 is at most three, which is the same as rolling doubles for the first time on the first, second, or third try. This should look familiar; it is the same as part (b). (In this context less than 4 and at most three are identical.) Therefore the answer to the third question is $1 - 0.4213 = 0.5787$

On the one hand, from part (b) the probability of rolling doubles in less than four tries is 0.4213 and it is rather likely that one of four players will get started quickly. On the other hand, from part (c) the probaility that it takes more than three turns is 0.5787 and each player should expect to take more than three turns in more than half the games played. □

In the next two sample problems we take a different tack. Rather than repeating the Bernoulli trial until the first success occurs, it will be repeated a fixed predetermined number of times usually denoted by n. So the number of successes (or failures) can only range from 0 to n. Then the natural question to ask is, what is the probability that a specified number of successes will occur? Of course, the specified number k of successes be must among the numbers $0, 1, 2, \ldots, n$.

Sample Problem 2. *In the course of a morning an interviewer for a pollster must ask six people to participate in an extensive public opinion survey. The interviewer knows that every time he or she asks to interview someone, the probability of being refused is 1/4.*

(a) What is the probability that all six people will agree to be interviewed?

(b) What is the probability that only the first and sixth will agree to be interviewed?

(c) What is the probability that exactly two of the six will agree to be interviewed?

SOLUTION: Attempting to interview someone should be viewed as an independent Bernoulli trial with failure being a refusal. Then the probability of success $p = 3/4$, the probability of failure $q = 1/4$, and the Bernoulli trial is repeated 6 times. As above A_i will denote success in the ith try. Thus we have six independent events - $A_1, A_2, A_3, A_4, A_5, A_6$.

(a) The probability that all six agree to be interviewed is the same as the probability of six successes or

$$A_1 \cap A_2 \cap A_3 \cap A_4 \cap A_5 \cap A_6.$$

Because these events are independent

$$\begin{aligned} P(A_1 \cap A_2 \cap A_3 \cap A_4 \cap A_5 \cap A_6) &= \frac{3}{4} \cdot \frac{3}{4} \cdot \frac{3}{4} \cdot \frac{3}{4} \cdot \frac{3}{4} \cdot \frac{3}{4} \\ &= \left(\frac{3}{4}\right)^6 = 0.178 \end{aligned}$$

(b) Here we must compute the probability of a success followed by 4 failures and a final success or

$$\begin{aligned} P(A_1 \cap A_2' \cap A_3' \cap A_4' \cap A_5' \cap A_6) &= \frac{3}{4} \cdot \frac{1}{4} \cdot \frac{1}{4} \cdot \frac{1}{4} \cdot \frac{1}{4} \cdot \frac{3}{4} \\ &= \left(\frac{1}{4}\right)^4 \cdot \left(\frac{3}{4}\right)^2 = 0.0022 \end{aligned}$$

(c) If we knew which of the 6 tries were to be successful, we could reason as in the previous parts. For example, the probability of 2 failures followed by 2 successes followed by 2 failures is

$$P(A_1' \cap A_2' \cap A_3 \cap A_4 \cap A_5' \cap A_6') = \left(\frac{3}{4}\right)^2 \cdot \left(\frac{1}{4}\right)^4 = 0.0022.$$

This answer is correct for any 2 specific successful trials, but there are $\binom{6}{2} = 15$ ways to pick which two tries will be successful. Therefore the answer is

$$15 \cdot (3/4)^2 \cdot (1/4)^4 = 0.0330.$$

The difference between the last part and the first two parts is that the event in the last part must be partitioned into 15 events like the one in part (b). □

The reasoning just used to complete the solution of the above example applies to any independent Bernoulli trial which is repeated n times. The probability of success in k specific tries and failure in the remaining $n-k$ tries equals $p^k \cdot q^{n-k}$ because the trials are independent. Since there are $\binom{n}{k}$ ways to select which k of the n tries will be successful, the probability of exactly k successes in n tries of an independent Bernoulli trial is

$$\binom{n}{k} \cdot p^k \cdot q^{n-k}$$

Sample Problem 3. *A typical roulette wheel in a casino has 38 numbered sectors and each number is equally likely. Two of the 38 sectors* *are colored green and marked 0 and 00 respectively. The remaining sectors are numbered from 1 to 36 with half red and the other half black. There is a whole range of allowed bets one of them being that the wheel will stop at a red number. Suppose a player bets on 10 consecutive spins of a roulette wheel that the outcome is a red number.*

(a) What is the probability that the player wins exactly 4 times?

(b) What is the probability that the player wins at most 2 times?

SOLUTION: In this situation success is the roulette wheel stopping at a red number and $p = 18/38 = 9/19$. This Bernoulli trial is repeated 10 times so in the above formula $n = 10$ and $q = 10/19$.

(a) The probability that the player wins exactly 4 times is given by the formula with $k = 4$. Substituting and using a calculator yields

$$\binom{10}{4} \cdot \left(\frac{9}{19}\right)^4 \cdot \left(\frac{10}{19}\right)^6 = 210 \cdot (0.4737)^4 \cdot (0.5263)^6 = 0.2247$$

(b) Winning at most 2 times is naturally partitioned into the three events - losing every time, winning exactly once, and winning exactly twice. The probabilities of each of these events can be calculated using the formula as in part (a) and the probabilities added to get the final answer. The probabilities of these events are:

$$\binom{10}{0} \cdot \left(\frac{10}{19}\right)^{10} = (0.5263)^{10} = 0.0016$$

$$\binom{10}{1} \cdot \left(\frac{9}{19}\right)^1 \cdot \left(\frac{10}{19}\right)^9 = 10 \cdot 0.4737 \cdot (0.5263)^9 = 0.0147$$

$$\binom{10}{2} \cdot \left(\frac{9}{19}\right)^2 \cdot \left(\frac{10}{19}\right)^8 = 45 \cdot (0.4737)^2 \cdot (0.5263)^8 = 0.0595$$

and their sum is 0.0758 which is the probability of winning at most two times out of ten tries.

Once again the key to solving the problem is recognizing when to partition an event into simpler events whose probabilities can be readily calculated using known formulas. □

Exercises

1. If the probability of success in a Bernoulli trial is 2/5 and the trial is repeated 45 times, what is the probability of exactly 10 successes?

2. Suppose E and F are independent events in a sample space with $P(E) = 1/3$ and $P(E \cap F) = 1/12$. Determine the following: a) $P(F)$, b) $P(E' \cap F)$, c) $P(E' \cap F')$, d) $P(E' \cup F')$.

3. If a fair coin is tossed 7 times, what is the probability of getting exactly 2 heads? At most 2 heads?

4. A fair die is rolled 15 times. What is the probability of getting a number less than 3 exactly seven times? What is the probability of getting a number less than 3 at least fourteen times?

5. Each weekday, the probability of finding a parking space in Lot 1 at 10:00 a.m. is .9. What is the probability that during one school week (Monday through Friday) you will find a parking space in Lot 1 at 10:00 a.m. exactly once?

6. A baseball player's batting average is .300. Assuming that this is his probability of getting a hit each time at bat, what is the probability that on his next 3 times at bat he will not get a hit but on his fourth time at bat he will get a hit? What is the probability he will get exactly 2 hits in his next 5 times at bat?

7. A certain Metro bus is supposed to arrive at the Student Union at 4:10 p.m. every weekday. The probability that it will be more than 5 minutes late is 1/3.

   a) What is the probability that out of the 21 weekdays in March, it will be more than 5 minutes late exactly 12 of the days?

   b) What is the probability that, during the same period, it will be more than 5 minutes late at least 2 of the days?

8. One of five dice lying on a table is selected at random and rolled 4 times. Only four of the five dice are fair; the event of rolling a 1 or 6 with the loaded die is 2/3. Given that the outcome was 1 or 6 exactly 3 times, what is the probability that the loaded die was selected?

3.5 Review Problems

1. An urn contains 5 blue balls and 1 yellow ball, a second urn contains 12 blue and 6 yellow balls, and a third urn contains 2 blue and 10 yellow balls. One of these three urns is selected at random and a ball is selected at random from it. What is the probability that a blue ball is selected? What is the probability that the ball selected comes from the first or second urn and is yellow? Given that a yellow ball is selected, what is the probability that it came from the third urn?

2. A pair of fair dice are rolled 100 times. What is the probability of not getting a six on either die at most 96 times?

3. Let M and N be events in a sample space S with $P(M) = 2/3$, $P(N) = 1/4$, and $P(M|N) = 1/3$. Compute $P(M \cup N)$.

4. When your TV breaks, three friends offer to fix it for free. The probabilities that you ask each of these friends to fix it are 1/6, 1/3, and 1/2 respectively. Unbeknown to you, the probabilities of their being able to fix it are 3/4, 3/8, and 5/8 respectively. What is the probability that your TV will be fixed? Given that one of them successfully repairs it, what is the probability that it was not the first friend?

5. A certain student has both a zoology and mathematics exam on the same day. The probability that the mathematics exam will be easy is .2 and the probability that the zoology exam will be easy is .3.

   (a) What is the probability that both exams will be easy?

   (b) What is the probability that at least one exam will be easy?

   (c) What is the probabililty that the mathematics exam will not be easy and the zoology exam will be easy?

6. Three coins are randomly selected at once from a dish containing 6 pennies and 4 dimes. Compute the probability of selecting at least one dime, given that at least one penny was selected.

7. Suppose D, E, and F are independent events in a sample space S with $P(D) = 1/2$, $P(D \cap E) = 1/6$, and $P(D \cap E \cap F) = 1/30$. Compute $P(E)$, $P(F)$, and $P(D \cup E \cup F)$.

8. The usual probability that an individual will contract the flu during the winter flu season is 1/25. If 3/4 of the population is given an experimental vaccine for the flu, and the probability that a vaccinated person will get it is 1/100, what is the probability that a person selected at random will catch the flu? Given that an individual gets the flu, what is the probability that he was vaccinated against it?

9. The probability that Doctor X will not be called back to the hospital on a Friday evening is 2/5.

   (a) In the course of a year (52 Fridays), what is the probability that he will not be called back exactly 30 Friday evenings?

   (b) What is the probability that during the same period there will be at least 50 Friday evenings when he is not called back?

10. A fair coin is tossed. If it comes up heads a number from 1 to 10 inclusive is selected at random; if it comes up tails a number from 1 to 20 inclusive is selected at random. Given that the number selected is less than 6, what is the probability that the outcome of the coin toss was heads?

11. Let E, F, and G be events in a sample space S with $P(E) = .3$, $P(F) = .4$, $P(G) = .5$, and $P(E \cap F) = .05$. Furthermore, E and G are disjoint and F and G are independent. Make a Venn diagram for E, F, and G and compute the probability of each cell.

12. A student has an important exam at 9:00 a.m.. Because he has an unreliable alarm clock, he asks two friends who don't know each other to call at 8:00 a.m. to wake him. He also sets his alarm, which has a probability of .4 of working any time it is set. The probabilities that his friends remember to call him are .6 and .7

respectively. What is the probability that the alarm will not work and both friends will forget to call? What is the probability that the alarm will not work and exactly one friend will remember to call him?

13. Suppose A and B are events in a sample space S with $P(A) = 0.3$, $P(B) = 0.2$, and $P(B|A) = 0.6$. Find $P(A|B)$.

14. A data processing company maintains 3 completely separate computers for customer use. At any given time the probabilities that they are not in service are 1/10, 1/5, and 1/2 respectively. When a customer tries to use these computers, what is the probability that he will find that:

    (a) All three computers are operating;

    (b) Exactly one of the computers is not in service;

    (c) At least two computers are not in service?

15. A sample space contains the events A and B satisfying $P(A) = 5/12$, $P(B) = 1/2$, and $P(A \cap B') = 1/3$.

    (a) Find $P(A \cap B)$.

    (b) Determine whether or not A and B are independent.

    (c) Compute $P(A'|B)$.

16. The probability that cookies dispensed by a vending machine are broken is 2/3. If you buy one pack of cookies every day for a week (7 days) from a vending machine, what is the probability of getting broken cookies exactly five of the days? What is the probability of getting broken cookies at most two of the days?

17. Three roommates plan to prepare desserts at home to bring back to the dorm after spring break. The first will bake a cake; unfortunately the chance his cakes are edible is only 1/4. The second will make fudge; unfortunately the chance his fudge is edible is only 2/5. The third will bake cookies; unfortunately the chance his cookies are edible is only 1/3. What is the chance that all the desserts will be edible? What is the chance that at least one of them will be edible?

18. A box contains some tickets and each of these tickets has a different three digit number printed on it. A ticket is selected at random. Let B be the event the number on the ticket selected begins with a 1 and let E be the event the number on the ticket selected ends with a 5. Determine whether or not B and E are independent for each of the following conditions on the tickets in the box:

    (a) The tickets are numbered with all the three digit numbers which can be formed using the digits 1, 2, 3, 4, and 5.

    (b) The tickets are numbered with all the three digit numbers formed using 1, 2, 3, 4, and 5 in which no digit is repeated.

19. Four urns contain the following mixtures of red and green balls: The first urn contains 6 red and 10 green balls, the second contains 15 red balls and 1 green ball, the third contains 2 red and 14 green balls, and the fourth contains 11 red and 5 green balls. An urn is selected at random and a ball is selected at random from that urn. What is the probability that a red ball is selected? Given that a green ball was selected, what is the probability that it came from the second urn?

20. Suppose A and B are events in a sample space S with $P(A) = 3/4$, $P(B') = 3/5$, and $P(A' \cap B') = 3/20$. Determine whether or not A and B are independent events.

21. The probability that the daily Broadway Special from Chicago to New York will arrive at least one hour late is 1/4.

    (a) What is the probability that during the month of February (28 days) it will be at least one hour late on exactly 15 days?

    (b) What is the probability that it will be "on time" (less than one hour late) on at least two days in February?

22. Let E and F be events in a sample space S with $P(E) = .4$ and $P(F|E) = 1$. Find $P(E \cap F)$.

23. A company has three machines A, B, and C, all producing the same part. Machine A produces 1/2 of the parts, machine B produces 1/4 of the parts, and machine C produces 1/4 of the parts. Furthermore 3/8 of the parts produced by A are defective, 1/4 of those produced by B are defective, and 3/4 of those produced by C are defective. What is the probability that a part selected at random will be defective? Given that the part is defective, what is the probability that it was made by machine C?

24. If E, F and G are three independnt events wit $P(E \cap F) = 1/6$ and $P(E \cap F \cap G) = 1/24$, find $P(G)$.

25. The Thompson's sick cat is taken to the new vet. The probability that the vet will recognize the cat's ailment and prescribe the right medicine is estimated to be .7. However, even if the right medicine is prescribed the cat has a probability of .2 of dying. But if a wrong medicine is prescribed this probability is .9. If the cat dies, what is the probability that the vet prescribed the right medicine?

26. Let A and B be independent events with $P(A) = 1/3$ and $P(B) = 1/4$. Determine $P(A'|B')$ and $P(A \cup B')$.

27. A marketing research report predicts that of the potential buyers of a certain product 5% will see the magazine advertisement for the product, 26% will see the television commercial for the product, and 1% will see both. Suppose 20% of those who see only the magazine advertisement will buy the product, 30% of those who see only the television commercial will buy it, 40% of those who see both will buy it, and 10% of those who see neither will buy it. What is the probability that a potential buyer selected at random will buy this product?

28. Data on smoking and lung cancer is displayed in the following table as percentages of the adult population:

	Smoker	Non-smoker
Develop lung cancer	4%	1%
No lung cancer	26%	69%

Find the probability that a smoker will develop lung cancer. How many times more likely is a smoker to develop lung cancer than a non-smoker?

29. An urn contains 3 white and 7 red balls; three balls are drawn in succession, one after another. When a white ball is drawn it is returned to the urn and when a red ball is drawn it is laid aside. What is the probability that the third ball drawn is white? (Hint: You will need a bigger tree diagram than usual.)

30. The game begins with the shooter rolling a pair of dice. The shooter wins with 7 or 11 (called a natural), loses with 2, 3, or 12 (called craps), and the remaining possibilities - 4, 5, 6, 8, 9, and 10 - become the shooter's point for the rest of the game. (In this context all the numbers naturally refer to the sum of the outcomes of the two dice.) If the shooter makes a point on the first roll, then the shooter wins by rolling the same point before rolling a 7 and loses by rolling a 7 before the point. (Casino rules are slightly different to ensure a take by the house on both bets with or against the shooter. Make a tree diagram (with many branches) for the shooter winning or losing. Use the tree diagram to calculate the probability that the shooter wins the game. If you are street smart do you bet for or against the shooter winning?

31. Diagnostic tests are rarely perfect and false positives and false negatives are a fact of life. Suppose a laboratory test to determine whether or not an individual is carrying a certain virus gives false positives and false negatives with probabilities 0.0005 and 0.01. If the probability that a random individual is carrying this virus is 0.001, what is the probability that given a positive test result the person tested is not carrying the virus?

32. The following is a direct quote from a letter by Steven Goldberg to the Editor of the New York Times in March 1994:

> "William Safire makes some powerful arguments against the lie detector in 'Holy Moley!' (column Feb 24) Two other arguments are compelling.

> (1) Under even the best of conditions , the lie detector incorrectly brands as liars a clearly unacceptable number of people who are telling the truth.
>
> Let us assume that a lie detector has an accuracy of 80 per cent (a generous estimate) and that no subject knows how to beat the lie detector. Let us further assume that 900 of a thousand people are telling the truth."

Answer the following questions which are discussed in the next two paragraphs of the letter:

(a) What is the probability that a person tells the truth and the lie detector identifies him or her as a liar?

(b) What is the probability that a person is identified as a liar by the lie detector?

(c) Given that a person was identified by the lie detector as a liar, what is the probability that they told the truth?

Chapter 4

Discrete Random Variables

Suppose a thousand people have been selected as a sample for some statistical study. These people can be thought of as the elements in a sample space. By asking each person in this sample space how many cars he or she has owned in his or her lifetime, we assign a number to each element in the sample space. Such an assignment of numbers to elements in a sample space will be called a random variable. Since there are many numerical facts such as age, salary, and height associated with each person's life, there are many different random variables for this sample space. In fact, on any sample space there are infinitely many random variables, but usually only a few of them will be of interest in a given situation.

At first glance it might appear that the use of random variables could be avoided by constructing sample spaces which contain only the information of interest. Unfortunately, decreasing the amount of information in the sample space usually makes it more difficult to compute the probabilities of events. For example, in most games which use two dice only the total number of dots on the top of the two dice is relevant. However, if we try to use the sample space consisting of the integers from 2 to 12, simple events are not equally likely and to calculate the probability of an event we would have to use the usual 36 element sample space. Thus it is easier to use the bigger sample space and think of the sum of the outcomes on the individual dice as a random variable. In fact, whenever probabilities are easily calculated in a sample space, the probabilistic behavior of any observed number or random variable can be determined and relevant calculations can be made.

The fundamental idea of probability is to predict with considerable certainty the average behavior of a random event repeated many times. A single value of a random variable is entirely unpredictable, but the expected value of a random variable provides a very reliable forecast of its average behavior. The profitability of casinos is ample evidence of the certainty with which the expected value predicts the long term behavior of a random variable. Although a casino can win or lose money on each spin of the roulette wheel, knowing that they have an expected gain of only a few cents on each dollar bet is enough to ensure their profit with a high volume of bets.

How accurately does the expected value predict the behavior of a random variable? The answer varies from one random variable to another, and the variance of a random variable provides a measure of the expected value's accuracy. A small variance means that the values of the random variable tend to cluster and do not exhibit a large deviation from the expected value. These two ideas, expected value and variance, are the primary topics of this section and play key roles in Chapters 5 and 6.

4.1 The Probability Function

Let S be a sample space. A *random variable* is a rule which assigns to each element in S a real number. Random variables will be denoted by capital letters like X, Y, and Z. The numbers which a random variable X assigns to elements of S are called the values of X. In this chapter we will only consider random variables whose values can be arranged in a finite or infinite list. Such random variables are called discrete. If X is a discrete random variable then the values of X will be denoted by x_1, x_2, x_3, etc.. For convenience we will also restrict our attention to integer valued discrete random variables.

For each value x_i of the random variable X, there must be some subset A_i of S consisting of those points in S to which X assigns the value x_i. Because X assigns exactly one value to each element of S, $A_i \cap A_j = \emptyset$ for $i \neq j$; that is, the events $A_1, \ldots, A_n$ are mutually exclusive. Since X assigns a value to each and every point of S, each point of S is in some A_i,

$$S = A_1 \cup A_2 \cup \ldots \cup A_n,$$

and $A_1, A_2, \ldots, A_n$ exhaust S. Thus every random variable X produces a partition of S. The probability that X equals x_i is just the probability of the event A_i, but we usually write $P(X = x_i)$ instead of $P(A_i)$. This procedure assigns to each value x_i of X the probability $P(X = x_i)$ which is called the *probability function* of X. Since $A_1, A_2, \ldots, A_n$ is a partition of S,

$$P(A_1) + P(A_2) + \ldots + P(A_n) = 1$$

or

$$P(X = x_1) + P(X = x_2) + \ldots + P(X = x_n) = 1.$$

We have already encountered several random variables and in principle know their probability functions; the only difference has been the absence of the standard terminology that was just introduced. Most of this section is devoted to understanding these earlier examples as random variables which is one of the fundamental concepts of probability. We begin with the familiar example of rolling a pair of dice and adding the two results to get an integer from 2 to 12.

Sample Problem 1. *Let S be the usual 36 element sample space used to describe the rolling of two fair dice. Let X be the sum of the outcomes of the individual dice, e.g. X assigns the number 8 to $(3,5)$. List the values of X, find the partition of S produced by X, and compute the probability function of X.*

SOLUTION: For convenience the sample space is reproduced again.

$$S = \left\{ \begin{array}{l} (1,1),(1,2),(1,3),(1,4),(1,5),(1,6), \\ (2,1),(2,2),(2,3),(2,4),(2,5),(2,6), \\ (3,1),(3,2),(3,3),(3,4),(3,5),(3,6), \\ (4,1),(4,2),(4,3),(4,4),(4,5),(4,6), \\ (5,1),(5,2),(5,3),(5,4),(5,5),(5,6), \\ (6,1),(6,2),(6,3),(6,4),(6,5),(6,6) \end{array} \right\}.$$

The values of X are clearly 2, 3, 4, 5, 6, 7, 8, 9, 10, 11, and 12. With the elements of S arranged as above in a 6 by 6 square, it is easy to find the partition produced by X. The random variable X assigns the same value to all the points on each diagonal line which is low on the left and high on the right. So X assigns the value

$$\begin{array}{rll}
2 & \text{to each point in} & \{(1,1)\}, \\
3 & \text{to each point in} & \{(2,1),(1,2)\}, \\
4 & \text{to each point in} & \{(3,1),(2,2),(1,3)\}, \\
5 & \text{to each point in} & \{(4,1),(3,2),(2,3),(1,4)\}, \\
6 & \text{to each point in} & \{(5,1),(4,2),(3,3),(2,4),(1,5)\}, \\
7 & \text{to each point in} & \{(6,1),(5,2),(4,3),(3,4),(2,5),(1,6)\}, \\
8 & \text{to each point in} & \{(6,2),(5,3),(4,4),(3,5),(2,6)\}, \\
9 & \text{to each point in} & \{(6,3),(5,4),(4,5),(3,6)\}, \\
10 & \text{to each point in} & \{(6,4),(5,5),(4,6)\}, \\
11 & \text{to each point in} & \{(6,5),(5,6)\}, \\
12 & \text{to each point in} & \{(6,6)\}.
\end{array}$$

In the above display the numbers on the left are the values of X and the sets on the right form the partition of S.

The last step is to calculate the probabilities of each of the events in this list. Putting together the values of the random variable and these probabilities, we have

$$\begin{array}{ll} P(X=2)=1/36, & P(X=3)=2/36, \\ P(X=4)=3/36, & P(X=5)=4/36, \\ P(X=6)=5/36, & P(X=7)=6/36, \\ P(X=8)=5/36, & P(X=9)=4/36, \\ P(X=10)=3/36, & P(X=11)=2/36 \\ P(X=12)=1/36. & \end{array}$$

which is the probability function of X. □

Sample Problem 2. *The probability that a certain basketball player can sink a foul shot is* $2/5$ *or* 0.4. *Suppose this player is allowed to shoot fouls until he or she makes one and let* X *be the number of tries it takes to do this. Determine the values of* X *and the probability function of* X.

SOLUTION: Clearly the values of X are $1, 2, 3, 4, \ldots, n, \ldots$; that is, all positive integers. This is an example of a discrete random variable whose list of values is infinite.

To calculate the probability function we will have to assume each shot is an independent Bernoulli trial with probability of success 2/5. As in Section 3.4 a Bernoulli trial is being repeated independently until the first success occurs, but now the number of tries until the first success is being counted and called the random variable X. So $P(X = 1)$ means the probability of success on the first try and must be 2/5. Next $P(X = 2)$ means the probability of missing the first shot and making the second. By the independence of the Bernoulli trial $P(X = 2) = (3/5)(2/5)$. Continuing in this way, $P(X = 3) = (3/5)^2(2/5)$. Thus

$$\begin{aligned} P(X=1) &= 2/5 = 0.4 \\ P(X=2) &= (3/5)\cdot(2/5) = 0.24 \\ P(X=3) &= (3/5)^2\cdot(2/5) = 0.144 \\ P(X=4) &= (3/5)^3\cdot(2/5) = 0.0864 \\ P(X=5) &= (3/5)^4\cdot(2/5) = 0.05184 \end{aligned}$$

and the general term is

$$P(X = k) = (3/5)^{k-1} \cdot (2/5).$$

The right hand is just a special case, $p = 2/5$ and $q = 3/5$, of the expression

$$q^{k-1} \cdot p$$

from Section 3.4. □

The previous example is just a specific illustration of what is called a geometric random variable. (The reason for the name is that the terms $P(X = n)$ form a geometric series because the ratio $P(X = n+1)/P(X = n)$ is a constant.) A *geometric random variable* is the number of times an independent Bernoulli trial must be repeated until the first success occurs. The values of a geometric random variable are always the positive integers. If as usual the probability of success is denoted by p and $q = 1 - p$, then

$$\begin{aligned} P(X = 1) &= p \\ P(X = 2) &= q \cdot p \\ P(X = 3) &= q^2 \cdot p \\ P(X = 4) &= q^3 \cdot p \end{aligned}$$

and the general term is

$$\boxed{P(X = k) = q^{k-1} \cdot p.}$$

Also in Section 3.4 on Bernoulli trials the probability of a specified number of successes in a fixed number of trials was determined. Counting the number of successes assigns an integer to the result of the experiment and defines another type of discrete random variable. In fact, determining the probability of exactly k successes in n tries amounts to calculating the probability function of a random variable.

To be more precise a *binomial random variable* X is the number of successes in n independent tries of a Bernoulli trial with probability of success equal to p. Clearly the values of X are $0, 1, 2, \ldots, n$ and from the last section of Chapter 3 we know that

$$P(X = k) = \binom{n}{k} \cdot p^k \cdot q^{n-k}.$$

A complete discussion of binomial random variables will be put off until Chapter 6 which is devoted entirely to this particularly important class of random variables.

A third common kind of discrete random variable is the hypergeometric random variable. Suppose r objects are randomly selected at once from n objects and the n objects fall into two disjoint classes - good and bad. A *hypergeometric random variable* X is the number of good objects selected. In particular, $P(X = k)$ is the probability of selecting exactly k good objects, which was determined in Section 3 of Chapter 2. With g and b standing for the number of good and bad objects respectively

$$\boxed{P(X = k) = \frac{\binom{g}{k}\binom{b}{r-k}}{\binom{n}{r}}.}$$

Sample Problem 3. *A contractor orders 10 truck loads of ready mixed concrete for a particular job. A sample is taken and tested from 3 of the 10 trucks. Let X be the number of samples which meet the contractor's specifications. Compute the probability function of X in the case that 4 of the 10 truck loads do not meet the contractor's specifications; for example, they might contain too much sand.*

SOLUTION: Clearly X is a hypergeometric random variable with values 0, 1, 2, and 3. Furthermore, $n = 10, g = 6, b = 4$, and $r = 3$. In the calculation of each $P(X = k)$ there will be a $\binom{10}{3}$ in the denominator, so we will do this calculation first.

$$\binom{10}{3} = \frac{10!}{3!7!} = \frac{8 \cdot 9 \cdot 10}{2 \cdot 3} = 4 \cdot 3 \cdot 10 = 120$$

The probability function is now calculated as follows:

$$
\begin{aligned}
P(X=0) &= \frac{\binom{6}{0}\binom{4}{3}}{\binom{10}{3}} = \frac{4}{120} = \frac{1}{30} = 0.0333 \\
P(X=1) &= \frac{\binom{6}{1}\binom{4}{2}}{\binom{10}{3}} = \frac{6 \cdot 6}{120} = \frac{36}{120} = \frac{3}{10} = 0.3 \\
P(X=2) &= \frac{\binom{6}{2}\binom{4}{1}}{\binom{10}{3}} = \frac{15 \cdot 4}{120} = \frac{60}{120} = \frac{1}{2} = 0.5 \\
P(X=3) &= \frac{\binom{6}{3}\binom{4}{0}}{\binom{10}{3}} = \frac{20}{120} = \frac{1}{6} = 0.1667.
\end{aligned}
$$

Although the above table contains complete information about the probability of finding 0, 1, 2, or 3 bad truck loads by testing 4 of the 10, it does not answer the question of when the results of the test lead the contractor to take action against the concrete supplier; that is a matter of human judgement. □

Pictorial devices often help understand, remember, and use mathematical ideas to wit Venn diagrams and tree diagrams. A histogram provides a way of picturing the probability function of a random variable. The key idea is to use area to represent probability.

If the dimensions of the base of a rectangle is one unit, then the area, which equals height times base, must be the same as the height. In what follows we will use rectangles with base one unit and height - hence also area - equal to $P(X = k)$, where X is an integer valued random variable.

To construct a *histogram* for X start by marking the integer values of X on a number line. For each value k draw a rectangle above the number line using the interval from $k-0.5$ to $k+0.5$ as its bottom. Use $P(X = k)$ for the height of the rectangle. Since the length of the base is $k + 0.5 - (k - 0.5) = k + 0.5 - k + 0.5 = 1$, the area of the rectangle is also $P(X = k)$. The drawing of all these rectangles together forms the histogram of X.

EXAMPLE 1. Let X be a discrete random variable with probability function $P(X = 1) = 0.1$, $P(X = 2) = 0.4$, $P(X = 3) = 0.2$ and $P(X = 4) = 0.3$. The histogram for X is shown in Figure 1 with the rectangle whose area equals $P(X = 2)$ shaded.

Sample Problem 4. *Construct a histogram for the hypergeometric random variable in Sample Problem 3.*

SOLUTION: The first steps are to list the values of the random variable and calculate its probability function. For convenience we chose an X for which this work had just been completed. From the solution to Sample Problem 3 we know that the values of X are $0, 1, 2, 3$ and the probability function is:

$$P(X = 0) = 0.0333 \qquad P(X = 1) = 0.3000$$
$$P(X = 2) = 0.5000 \qquad P(X = 3) = 0.1667$$

Since there are four possible values for X, we must draw four rectangles. The first one for the value 0 will be over the interval from -0.5 to 0.5 and will be very small with a height of only 0.0333. The next rectangle will have its base from 0.5 to 1.5 and a height of $P(X = 1) = 0.3$. The other two rectangles are drawn in like manner and the entire histogram is shown in Figure 2. □

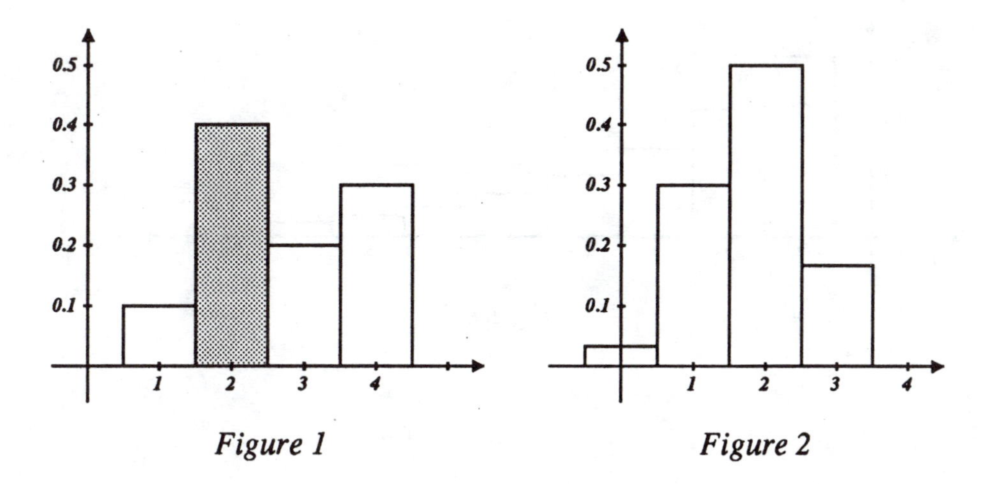

Figure 1 *Figure 2*

Sample Problem 5. *Construct a histogram for a geometric random variable X with $p = 0.3$ and $q = 0.7$.*

SOLUTION: The values of X are the positive integers, $1, 2, 3, \ldots$. Since X has infinitely many values it is impossible to draw the entire histogram. However, drawing it for just the first few values will suffice to give us a mental picture of the behavior of its probability function.

Using the formula

$$P(X = k) = q^{k-1} \cdot p,$$

we find that

$$\begin{aligned}
P(X = 1) &= 0.3 \\
P(X = 2) &= 0.7 \cdot 0.3 = 0.21 \\
P(X = 3) &= (0.7)^2 \cdot 0.3 = 0.147 \\
P(X = 4) &= (0.7)^3 \cdot 0.3 = 0.103 \\
P(X = 5) &= (0.7)^4 \cdot 0.3 = 0.072 \\
P(X = 6) &= (0.7)^5 \cdot 0.3 = 0.050.
\end{aligned}$$

Now construct rectangles for these six values as in the previous Sample Problem. They are shown with a few more in Figure 3. Although Figure 3 is not the entire histogram for this geometric random variable, it clearly shows that the rectangles get lower and lower as the values of X get bigger. □

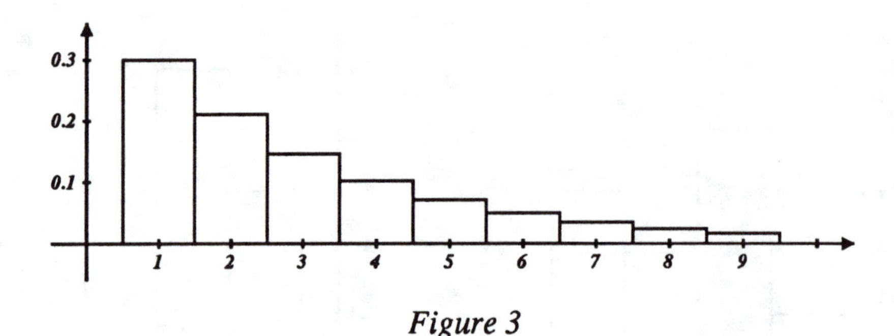

Figure 3

Exercises

1. Let X be the number of heads in four tosses of a fair coin. List the values of X, find the partition of the usual sample space S for four tosses of a coin, compute the probability function of X, and construct a histogram for X.

2. Two fair four-sided dice (tetrahedrons) are rolled. (As you would expect, the sides are marked with one, two, three, or four dots and the outcome of a single roll is the number of dots on the side face down.) Let X be the product of the outcomes of the individual dice. Compute the probability function of X and construct a histogram for X.

3. The faces of a fair twelve-sided (dodecahedron) die are numbered from 1 to 12. (This is one of the kinds of dice used in Dungeons and Dragons.) Let X be the remainder obtained from dividing the outcome of a roll of this die by 5. Compute the probability function of X.

4. Suppose $P(E) = 0.2, P(F) = 0.3$, and $P(E \cap F) = 0.05$ when E and F are events in S. Let X assign the value 0, 1, or 2 to a point s in S according as s is in neither E nor F, in exactly one of E or F, or in both E and F. Compute the probability function of X and make a Venn diagram for E and F in which each set of the partition associated with X is shaded differently.

5. The probability function of a random variable X is given by the following table:

$$\begin{array}{rcccccc} x_i & : & -1 & 2 & 5 & 6 & 9 \\ P(X = x_i) & : & .05 & .15 & a & .25 & .1. \end{array}$$

   Find $a = P(X = 5)$ and construct a histogram for X.

6. An electronic game asks the person playing it a trivia question and lists four possible answers of which only one is correct. If the player's answer is wrong, it gives the correct answer and moves on to the next question. Suppose a player answers by randomly

guessing. Let X be the number of questions this player answers until he answers one correctly. Compute $P(X=1)$, $P(X=2)$ and $P(X=3)$.

7. Let X be a geometric random variable coming from a Bernoulli trial with probability of success 2/3. Compute the probability that the value of X is less than 5 and the probability that the value of X is at least 3.

8. Let X be a random variable with values 1, 2, 3, and 4, and suppose $P(X=1)=1/c, P(X=2)=2/c, P(X=3)=3/c$, and $P(X=4)=4/c$. Find c.

9. Suppose you randomly select two ball point pens from the nine in your desk drawer, and suppose four of nine are no longer usable. Compute the probability function for X equals the number of usuable pens you select.

10. Let X be a hypergeometric variable with $n=11$, $r=4$, $g=6$. Compute the probability function of X and construct a histogram for X.

11. The following table gives the probability function of a random variable X:

$$\begin{array}{rcccccc} x_i & : & -4 & -2 & 0 & 1 & 3 \\ P(X=x_i) & : & 0.2 & 0.25 & 0.15 & 0.1 & 0.3. \end{array}$$

Compute the probabilities of the following events:

(a) the value of X is negative,

(b) the value of X is greater than -3,

(c) the value of X is at most 0.

12. A bowl contains 10 poker chips, 6 white chips worth one dollar each, 3 red chips worth five dollars each, and 1 blue chip worth ten dollars. Three chips are randomly selected at once. Let X be the total value of the chips selected. Compute the probability function of X and construct a histogram for X.

13. Three fair dice are rolled. Let Z be the sum of the three outcomes. What are the values of Z? Compute $P(Z = 6), P(Z = 7)$ and $P(Z = 8)$.

14. Four out of ten loaves of bread on a supermarket shelf are stale. You pick three loaves randomly. Let X indicate the number of fresh loaves selected. Find the probability function of X.

15. A box contains 8 light bulbs of which 2 are defective. Bulbs are drawn one at a time at random and discarded until a good bulb is drawn. Let X be the number of the draw of the first good bulb. Compute the probability function for X.

16. Construct a histogram for a geometric random variable with $p = 0.5$.

17. Construct a histogram for a binomial random variable with $p = 0.25$ and $n = 4$.

18. Construct a histogram for a hypergeometric random variable with $n = 9$, $g = 5$, and $r = 3$.

4.2 The Expected Value

Let X be a random variable; we would like to compute an average value for X. If $P(X = x_1)$ is much smaller then $P(X = x_2)$, then x_1 is not as significant a value of X as is x_2 and should not count as much as x_2 in an average.

As an illustration of how $P(X = x_i)$ enters the averaging process consider the following situation: In a school census 100 families are asked among other things how many children they have. So we have a random variable X which assigns to each family the number of children in the family. Suppose the values of X are 1, 2, 3, 4, 5, 6. Furthermore, suppose there are 35 families with 1 child, 40 families with 2 children, 15 families with 3 children, 5 families with 4 children, 3 families with 5 children, and 2 families with 6 children. If we simply compute the average of the values of X, we get

$$\frac{1+2+3+4+5+6}{6} = \frac{21}{6} = 3.5$$

which is obviously nonsense because only 10 families have more than 3 children. What we should have done was multiplied 1 by the number of families with 1 child, 2 by the number of families with 2 children and so forth and then divided by 100, not 6, as in:

$$\frac{1 \cdot 35 + 2 \cdot 40 + 3 \cdot 15 + 4 \cdot 5 + 5 \cdot 3 + 6 \cdot 2}{100} = 2.07$$

If we rewrite this expression as a sum of fractions then the probability function of X will appear. Specifically,

$$\frac{1 \cdot 35}{100} + \frac{2 \cdot 40}{100} + \frac{3 \cdot 15}{100} + \frac{4 \cdot 5}{100} + \frac{5 \cdot 3}{100} + \frac{6 \cdot 2}{100} =$$

$$1 \cdot \frac{35}{100} + 2 \cdot \frac{40}{100} + 3 \cdot \frac{15}{100} + 4 \cdot \frac{5}{100} + 5 \cdot \frac{3}{100} + 6 \cdot \frac{2}{100} = 2.07$$

The probability of randomly selecting a family from this sample with 1 child is 35/100 and

$$P(X = 1) = \frac{35}{100}.$$

Similarly,

$$P(X=2)=\tfrac{40}{100} \quad P(X=3)=\tfrac{15}{100}$$

$$P(X=4)=\tfrac{5}{100} \quad P(X=5)=\tfrac{3}{100}$$

$$P(X=6)=\tfrac{2}{100}$$

and the preceding equation is the same as

$$\begin{aligned} & 1\cdot P(X=1)+2\cdot P(X=2)+3\cdot P(X=3) \\ + \; & 4\cdot P(X=4)+5\cdot P(X=5)+6\cdot P(X=6) \\ = \; & 2.07 \end{aligned}$$

This last equation shows how averaging can be understood in terms of probabilities and is the basis for the next definition.

The *expected value* or *mean* of a random variable X with values $x_1, \ldots, x_n$ is defined by

$$\boxed{E[X] = x_1 \cdot P(X=x_1) + x_2 \cdot P(X=x_2) + \ldots + x_n \cdot P(X=x_n).}$$

Sample Problem 1. *A pair of fair dice is rolled. Let X be the sum of the outcomes of the individual dice, e.g. the value of X is 5 when you get a 2 on one die and a 3 on the other. Compute the expected value of X.*

SOLUTION: This is the same random variable whose values and probability function were determined in Sample Problem 1 of the previous section. In particular it was shown that

$$\begin{array}{ll} P(X=2)=1/36, & P(X=3)=2/36, \\ P(X=4)=3/36, & P(X=5)=4/36, \\ P(X=6)=5/36, & P(X=7)=6/36, \\ P(X=8)=5/36, & P(X=9)=4/36, \\ P(X=10)=3/36, & P(X=11)=2/36 \\ P(X=12)=1/36. & \end{array}$$

Using this information and the definition of $E[X]$,

$$E[X] = 2\cdot\frac{1}{36}+3\cdot\frac{2}{36}+4\cdot\frac{3}{36}+5\cdot\frac{4}{36}+6\cdot\frac{5}{36}$$

$$\begin{aligned}
&+ \quad 7\cdot\frac{6}{36}+8\cdot\frac{5}{36}+9\cdot\frac{4}{36}+10\cdot\frac{3}{36}+11\cdot\frac{2}{36}+12\cdot\frac{1}{36} \\
&= \quad \frac{2+6+12+20+30+42+40+36+30+22+12}{36} \\
&= \quad \frac{252}{36}=7
\end{aligned}$$

to complete the solution. □

Sample Problem 2. *A tire company promotes a protection plan insuring a new pair of its tires against blowouts for up to* 50,000 *miles. The cost is a one time premium of* \$10.00 *per pair of tires. The actual cost to the company for manufacturing, delivering, and installing one of these tires is* \$75.00. *They know from experience that within* 50,000 *miles the probability that a policy holder will have blowouts in both tires is* 0.01 *and the probability that a policy holder will have a blowout in one of the two insured tires is* 0.05. *What is the company's expected gain (or loss) on each policy sold?*

SOLUTION: Let X be the company's gain on a policy. (This is a random variable on the sample space consisting of the policies sold.) The problem is to compute $E[X]$. The values of X are $x_1 = 10$, $x_2 = -65$, and $x_3 = -140$. Notice that the premium the company collected was deducted from the cost of replacing the tire(s). We are given that $P(X = x_2) = 0.05$, $P(X = x_3) = 0.01$, so it follows that $P(X = x_1) = 0.94$. Finally

$$\begin{aligned}
E[X] &= x_1\cdot P(X=x_1)+x_2\cdot P(X=x_2)+x_3\cdot P(X=x_3) \\
&= 10\cdot 0.94+(-65)\cdot 0.05+(-140)\cdot 0.01 \\
&= 9.4-3.25-1.40=4.75.
\end{aligned}$$

which is a fantastic deal for the tire company. □

The expected value of a random variable X is frequently denoted by μ, the lower case Greek letter mu, instead of the more cumbersome

$E[X]$. There are also special formulas for the expected values of different kinds of random variables. For example, if X is the geometric random variable where the probability of success on one trial is p, then

$$\mu = E[X] = \frac{1}{p}.$$

This means that the smaller the p, the longer you can expect to wait for the first success.

EXAMPLE 1. Let X be the number of times a pair of dice is rolled until doubles appears for the first time. This is a geometric random variable with $p = 1/6$ and by the above formula.

$$\mu = E[X] = \frac{1}{\frac{1}{6}} = 6.$$

The formula for the mean of a hypergeometric random variable is

$$\mu = E[X] = \frac{g}{n} \cdot r$$

where g is the number of good objects, n is the total number of objects, r is the number selected, and $r \leq g$. The formula for the mean of the hypergeometric random variable is intuitively natural. Note that the probability of selecting one good object at random is $\frac{g}{n}$, the proportion of good objects. So when selecting r objects, we should expect the same proportion of good ones, that is

$$\frac{g}{n} \cdot r.$$

EXAMPLE 2. Suppose 6 apples are randomly selected from 80 in a grocery store display. If 20 of the 80 are wormy, how many good apples are expected in the 6 selected. Here $n = 80$, $g = 60$, and $r = 6$. So $g/n = 60/80 = 3/4$ and

$$\mu = \frac{3}{4} \cdot 6 = 4.5.$$

Let X be a random variable on the sample space S with values $x_1, \ldots, x_n$, and let a and b be two real numbers with $a \neq 0$. Then

$Y = aX + b$ is also a random variable, but with values $ax_1 + b$, $ax_2 + b$ etc.. Although the values of Y are different from those of X, Y produces the same partition on S because Y assigns the value $ax_i + b$ to the same points to which X assigns the value x_i. It follows that

$$P(Y = ax_i + b) = P(X = x_i),$$

and not surprisingly there is a simple linear relationship between the expected values of and X and Y, namely

$$\boxed{E[Y] = E[aX + b] = aE[X] + b}$$

EXAMPLE 3. Let X be a random variable whose values are in degrees celsius and whose expected value is 20. Then

$$Y = \frac{9}{5}X + 32$$

is the corresponding random variable whose values are in degrees Fahrenheit and

$$E[Y] = \frac{9}{5}E[X] + 32 = \frac{9}{5} \cdot 20 + 32 = 68.$$

Sample Problem 3. *Let X be a random variable with mean 7. Find the means of Y and Z if $Y = 2X + 3$ and $X = 5Z - 13$.*

SOLUTION: Remember mean is a synonym for expected value, so the problem is to compute $E[Y]$ and $E[Z]$. Using the above formula for Y,

$$E[Y] = E[2X + 3] = 2E[X] + 3 = 2 \cdot 7 + 3 = 17.$$

When the same formula is used on $X = 5Z - 13$,

$$E[X] = E[5Z - 13] = 5E[Z] - 13.$$

Since $E[X] = 7$ is given,

$$7 = 5E[Z] - 13$$

which can easily be solved to show that $E[Z] = 4$. □

The expected value of a random variable provides a useful mathematical definition of a fair game. Let X be the gain of a specific player in a game of chance on which some bets have been placed. The game is said to be *fair* if $E[X] = 0$. Furthermore, if $E[X] > 0$ we say the game favors the designated player, and if $E[X] < 0$ we say the game favors the opponent(s). It is important that the values of X be consistently determined from a specific players viewpoint and that negative numbers be used for losses.

Sample Problem 4. *Chuck-a-Luck is a common carnival dice game. The player picks a number from 1 to 6 and bets a dollar (or more) on it. Then three dice are rolled. The player wins a dollar for each occurrence of the number on which the bet was placed and otherwise loses the dollar bet. For example, if the player bets on sixes and the outcome is* $(6, 1, 6)$, *then the player wins two dollars. What is the player's expected gain on a dollar bet?*

SOLUTION: Let X be the players gain. Because it is a carnival game we should suspect that $E[X]$ will be negative, but many people reason that it is fair as follows: Since there is a one sixth chance of winning on each die and there are three dice, the chance of winning should be $3 \cdot \frac{1}{6} = \frac{1}{2} = 0.5$. The mistake here is adding the probabilities of events that are not disjoint; you can win on two or three dice at the same time. The real question here is, how good a money maker is Chuck-a-Luck for the carnival?

To make the analysis a little easier to follow let's continue to assume that the player bets on sixes. The values of X are clearly $-1, 1, 2, 3$. The value of X is -1 when no sixes occur and the player loses the dollar bet. There are 5 ways this can happen on each die and hence by independence,

$$P(X = -1) = \left(\frac{5}{6}\right)^3 = \frac{125}{216} = 0.5787.$$

The player wins 1 dollar when 1 six occurs. The lone six can occur on any one of the three dice and there are then 5 possibilities for the other two dice. It follows by the Multiplication Principle that

$$P(X = 1) = \binom{3}{1} \cdot \left(\frac{1}{6}\right) \cdot \left(\frac{5}{6}\right)^2 = \frac{75}{216} = 0.3472.$$

In like manner

$$P(X = 2) = \binom{3}{2} \cdot \left(\frac{1}{6}\right)^2 \cdot \left(\frac{5}{6}\right) = \frac{15}{216} = 0.0694$$

and

$$P(X = 3) = \left(\frac{1}{6}\right)^3 = \frac{1}{216} = 0.0046$$

(The reasoning here is no different than calculating the probability of k successes in 3 independent tries of a Bernoulli trial.) Putting these calculations together,

$$E[X] = -1 \cdot 0.5787 + 1 \cdot 0.3472 + 2 \cdot 0.0694 + 3 \cdot 0.0046 = -0.079$$

Looking at the game from the carnival's point of view they can expect to make almost eight cents on every dollar bet. □

In some games of chance the player must first pay a fee to play and then collects his winnings, if any, at the end of the game. This fee must be taken into account when calculating the player's gain. As before let X be the player's gain and now let W be the winnings or payoff the player receives when the game is over.

If f is the fee to play, then the relationship between X and W is

$$X = W - f.$$

Since $E[W - f] = E[W] - f$,

$$E[X] = E[W] - f.$$

Therefore, a game for which a fee f must be paid to play is fair provided

$$E[W] - f = 0$$

or more simply

$$E[W] = f.$$

Of course, such a game favors the player if $E[W] > f$ and the house if $E[W] < f$ which is the usual state of affairs.

Keno is a well known game which you pay a fee to play. It is found in casinos and more recently it has become part of some state lottery operations. To play Keno you fill out a play slip by picking a required number of integers from 1 to 80, turn it in to the person operating the game, and pay your fee for a ticket. Periodically the casino or lottery randomly selects twenty numbers at once from 1 to 80. Your winnings are based on how well the numbers you picked match theirs. The next Sample Problem looks at the Maryland 4 Spot Game.

Sample Problem 5. *In the Maryland Lottery's 4 Spot Game you pick four numbers, pay one dollar to play, and win* \$50.00, \$5.00, *or* \$2.00 *according as 4, 3, or 2 of your numbers match numbers in the twenty selected by the Maryland Lottery.*

(a) Calculate a player's expected winnings on a dollar bet.

(b) Calculate the Lottery's expected gain on a dollar bet.

SOLUTION: Let W denote the player's winnings on a dollar bet. It is a random variable with values 50, 5, and 2. The critical part of the problem is correctly calculating the probability function for W. The Lottery randomly picks 20 different numbers from 80 numbers Of these 80 numbers, 4 are the players numbers and 76 are not. So the numbers fall into two classes, 4 that are good for the player and 76 that are bad. Thus the probability of matching 4 numbers and winning fifty dollars is the same as the probability that the Lottery selects 4 good numbers and 16 bad numbers which is a hypergeometric probability. (The random varaible W is not itself a hypergeometric random variable because it does more than count the number of matches - it assigns prize money to some of them.) It follows that

$$P(W = 50) = \frac{\binom{4}{4} \cdot \binom{76}{16}}{\binom{80}{20}} = 0.0031.$$

The probabilities $P(W = 5)$ and $P(W = 2)$ are also hypergeometric and are done in the same way:

$$P(W = 5) = \frac{\binom{4}{3} \cdot \binom{76}{17}}{\binom{80}{20}} = 0.0432.$$

$$P(W = 2) = \frac{\binom{4}{2} \cdot \binom{76}{18}}{\binom{80}{20}} = 0.2126.$$

It is not necessary to compute $P(W = 0)$ because it will enter the calculation of $E[W]$ as $0 \cdot P(W = 0) = 0$.

(a) The player's expected winnings is $E[W]$ and using the above calculations

$$E[W] = 50 \cdot 0.0031 + 5 \cdot 0.0432 + 2 \cdot 0.2126 = 0.5836.$$

Hence the frequent player can expect to win about 58 cents on a dollar bet.

(b) The simplest way to calculate the Lottery's expected gain on a dollar bet is to express the Lottery's gain denoted by Y in terms of W. The Lottery collects the one dollar fee and pays W to the player leaving it with $Y = f - W = 1 - W$. Consequently

$$E[Y] = E[1 - W] = 1 - E[W] = 1 - 0.5836 = 0.4164$$

The other nine games in the Maryland Lottery's Keno operation can be analyzed in the same way. □

In summary, our analysis of games is based on *first* dividing them into two kinds - those in which no money changes hands until play is completed and those in which a fee must be paid before play begins. When one plays and then pays or collects, the appropriate random variable X is your gain and such a game is fair provided $E(X) = 0$. This category of games includes playing gin rummy or bridge for a penny a point, flipping a coin to decide who buys the coffee, and betting on a roulette wheel. When one must first pay a fee f before being allowed to play, it is more convenient to use your winnings W as the random variable. In this case the test for fairness is the equation $E(W) = f$. Bingo, raffles, slot machines, keno, and lotteries are all games which require a fee to play.

Exercises

1. If the probability function of X is $P(X = -2) = 1/7$, $P(X = 1) = 2/7$, $P(X = 3) = 3/7$, and $P(X = 5) = 1/7$, what is the expected value of X?

2. The following table gives the values and probabilities of a random variable X:

x_i	-4	-2	0	1	3
$P(X = x_i)$	0.2	0.25	0.15	0.1	0.3.

   Determine $E[X]$.

3. Let X be the number of times a pair of fair dice are rolled until the first doubles occur. Compute the mean of X.

4. Two doughnuts are selected at random from a dozen doughnuts of which 7 are jelly filled and 5 are creme filled. Let X be the number of jelly filled doughnuts selected. Find the mean of X.

5. The probability function of a random variable X is given by the following table:

x_i	-5	-3	-1	0	2	4
$P(X = x_i)$	1/10	2/10	1/10	3/10	1/10	2/10.

   Calculate the expectation of X.

6. A restaurant keeps track of how many one hundred pound bags of ice they use each week during the 10 week period of summer preceding Labor Day. There are 3 weeks when they use 7 bags, 1 week they use 8 bags, 3 weeks they use 9 bags, and 1 week each they use 10, 11 and 12 bags. What is the expected number of bags they use each week?

7. If W and Y are random variables such that $W = 3Y + 8$ and $E[W] = 20$, what is $E[Y]$?

8. Let X and Y be random variables with $E[X] = 10$ and $E[Y] = 3$. If $X = 7Y + C$, what does C equal?

9. Find $E[X]$ for the random variable X with probability function:

$$\begin{array}{rcrrrr} x_i & : & -2 & -1 & 1 & 2 \\ P(X = x_i) & : & 0.1 & 0.2 & 0.4 & 0.3. \end{array}$$

What does $E[10X + 7]$ equal?

10. Compute the expected values of the random variables in Exercises 2, 3, 4, 6, 9 and 12 from the previous section.

11. You and a friend gamble with a fair die. If you win \$1.00, \$2.00 or \$3.00 when the outcome of the roll of the die is 1, 2 or 3 respectively, what should your opponent win when the outcome is not 1, 2 or 3 to make the game fair?

12. A card is to be selected at random from a standard deck. Is the following a fair bet? If an ace is selected, you win \$2.60; otherwise you lose a quarter.

13. At a state fair there is a "spin the wheel game". The wheel is numbered from 1 to 100 and each number is equally likely. It costs \$1.00 to play the game. If the wheel stops at the number 50 you win \$50.00, if it stops at 25 you win \$25.00, and if it stops at 7 or 11 you win \$5.00. Whom does the game favor and by how much? What should it cost for this to be a fair game?

14. Suppose you play a game with a special deck of cards consisting of 4 white cards, 4 red cards, 4 blue cards, 4 yellow cards, and 2 black cards. You draw a card. If it is a white card you win \$5.00; if it is a red, blue or yellow card you win \$1.00. How much should you pay your opponent when you draw a black card, to make the game fair?

15. A grab bag contains 20 packages worth 7 dollars each, 20 packages worth 12 dollars each, and 5 packages worth 14 dollars each. What would be a fair price for the privilege of selecting one of these packages at random?

16. A casino game consists of a single toss of a fair die and pays off as follows: players win \$1.00 if they toss a "one", \$3.00 if they toss

a "three", and \$5.00 if they toss a "five". It costs a player \$2.00 to play (that is, for one toss of the die). Does this game favor the player or the casino? What should the casino charge to make it a fair game?

17. In a carnival game the customer throws 2 balls at 5 duck pins stacked in a pyramid. The probability of knocking down all 5 in 2 throws is .1, with a prize worth \$1.00; the probability of knocking down 4 pins in 2 throws is .2, with a prize worth 50 cents; the probability of knocking down 3 pins in 2 throws is .3, with a prize worth 20 cents. There is no prize for knocking down 2 pins or less. If the game costs 35 cents to play, how much profit can the carnival expect to make per game?

18. In a gambling game you roll two fair dice. You win \$4.00 for a 2, 7 or 12 and lose \$1.00 otherwise. Would you be willing to play this game repeatedly? Why or why not?

19. At a carnival you can play the following game for \$2.00: A bag contains 3 white balls and 1 red ball. From this bag you randomly select balls one at a time without replacement. If the red ball is the last one selected you win \$4.00, if it is the third ball selected you win \$2.00, if it is the second ball selected you win \$1.00 and if it is the first ball selected you win nothing. What is your expected gain for this game?

20. A typical roulette wheel in a casino has 38 numbered sectors and each number is equally likely. Two of the 38 sectors are colored green and marked 0 and 00 respectively. The remaining sectors are numbered from 1 to 36 with half red and the other half black. There is a whole range of allowed bets one of them being that the wheel will stop at a red number. For each of the following bets calculate a player's expected gain on a one dollar bet for the given house odds:

    (a) a red number at 1:1,

    (b) the first dozen (1 to 12) at 2:1,

    (c) a row (3 numbers in a row on the betting table) at 11:1,

(d) a single number at 35:1 .

21. Calculate a players expected gain on a one dollar bet on the Maryland Lottery's 3 Spot Game. In this game the player wins $25 if all three numbers match and $2 if two numbers match.

22. Calculate a player's expected gain on a one dollar bet on the Maryland Lottery's 5 spot game. In this game the player wins $300 if all five numbers match, $50 if four match, $15 if three match, and $2 if two match.

4.3 The Variance

Let X be a random variable with values $x_1, \ldots, x_n$. As in the previous section the expected value or mean will be denoted by μ. How closely can we expect the values of X to cluster around the mean μ ? The use of the word "expected" was deliberate; we would like to define some sort of expected value which would measure the dispersion of the values of X. This would provide a means of comparing the behavior of different random variables.

The simplest thing to try is

$$(x_1 - \mu)P(X = x_1) + \ldots + (x_n - \mu)P(X = x_n).$$

Because some of the values of X are less than μ and others are greater, the effect of a value of X much less than μ can be cancelled by value(s) bigger than μ. In fact, the above expression is always zero, that is

$$(x_1 - \mu)P(X = x_1) + \ldots + (x_n - \mu)P(X = x_n) = 0.$$

One easy way to get rid of the negative terms is to square all the $(x_i - \mu)$ terms.

The *variance* of X is defined by

$$V[X] = (x_1 - \mu)^2 P(X = x_1) + \ldots + (x_n - \mu)^2 P(X = x_n).$$

One disadvantage of squaring these terms is that all the units are squared. For example, if the values of X are costs in dollars, then the units for $V[X]$ are square dollars. To avoid this we define the *standard deviation* by

$$\sigma[X] = \sqrt{V[X]}.$$

(σ is the lower case Greek sigma.) The standard deviation is often denoted just by σ and the variance by σ^2.

Roughly speaking the larger the variance or standard deviation, the less the values of X will cluster close to the mean. When the variance is small there may be values of X at a considerable distance from μ , but their probabilities will be small so their contribution $(x_i - \mu)^2 P(X = x_i)$ to the variance is small. In other words, when the variance or standard deviation is small only the values close to μ occur frequently.

The formula for the variance can be simplified by using

$$(x_i - \mu)^2 = x_i^2 - 2x_i\mu + \mu^2.$$

The result of expanding each of the terms and collecting the terms is

$$\boxed{V[X] = E[X^2] - \mu^2}$$

which is usually the preferred method of computing $V[X]$. If X is a random variable, then squaring the value of X produces a new random variable denoted by X^2 with value $x_1^2, x_2^2, x_3^2, \ldots, x_n^2$. The formula $V[X] = E[X^2] - \mu^2$ requires the calculation of the expected value of this new random variable X^2 as well as that of the original random variable X.

Let X and Y be two random variables with the same values and probability functions are given by the following table:

Values	1	2	3	4	5
$P(X = x_i)$	1/8	3/16	3/8	3/16	1/8
$P(Y = y_i)$	1/16	1/4	3/8	1/4	1/16

These random variables were designed to illustrate the meaning of variance without any unnecessary complexity. Most random variables are not this simple.

First their expected values are equal as the following calculations show:

$$\begin{aligned} E[X] &= 1 \cdot \frac{1}{8} + 2 \cdot \frac{3}{16} + 3 \cdot \frac{3}{8} + 4 \cdot \frac{3}{16} + 5 \cdot \frac{1}{8} \\ &= \frac{2}{16} + \frac{6}{16} + \frac{18}{16} + \frac{12}{1} + \frac{10}{16} = \frac{48}{16} = 3 \end{aligned}$$

and

$$\begin{aligned} E[Y] &= 1 \cdot \frac{1}{16} + 2 \cdot \frac{1}{4} + 3 \cdot \frac{3}{8} + 4 \cdot \frac{1}{4} + 5 \cdot \frac{1}{16} \\ &= \frac{1}{16} + \frac{8}{16} + \frac{18}{16} + \frac{16}{16} + \frac{5}{16} = \frac{48}{16} = 3. \end{aligned}$$

The essential difference between X and Y is that the values 2 and 4 occur more frequently for Y than for X, and the values 1 and 5 occur less frequently for Y than for X. Since 2 and 4 are closer to their common mean of 3 than are 1 and 5, the variance of Y should be smaller than the variance of X.

The variances of X and Y will be calculated using the formula $V[X] = E[X^2] - \mu^2$. By squaring the values in the above tables giving the probability functions for X and Y produces the probability functions for X^2 and Y^2:

Values	1	4	9	16	25
$P(X = x_i^2)$	1/8	3/16	3/8	3/16	1/8
$P(Y = y_i^2)$	1/16	1/4	3/8	1/4	1/16

Next calculate the expected values of X^2 and Y^2 in the usual way as follows:

$$E[X^2] = 1 \cdot \frac{1}{8} + 4 \cdot \frac{3}{16} + 9 \cdot \frac{3}{8} + 16 \cdot \frac{3}{16} + 25 \cdot \frac{1}{8} = \frac{166}{16}$$

$$E[Y^2] = 1 \cdot \frac{1}{16} + 4 \cdot \frac{1}{4} + 9 \cdot \frac{3}{8} + 16 \cdot \frac{1}{4} + 25 \cdot \frac{1}{16} = \frac{160}{16} = 10$$

Finally

$$V[X] = \frac{166}{16} - 3^2 = \frac{166}{16} - 9 = \frac{166}{16} - \frac{144}{16} = \frac{22}{16} = 1.3750$$

$$V[Y] = 10 - 3^2 = 1.$$

and as predicted $V[Y]$ is less than $V[X]$. This comparison between the variances of X and Y is also evident in their histograms shown in Figures 1 and 2 respectively. Notice that in the histogram of Y (Figure 2) there is more of the area under the curve concentrated near the mean 3 than for X (Figure 1). Pictorially speaking the more widely the area under the histogram is spread the larger the variance is.

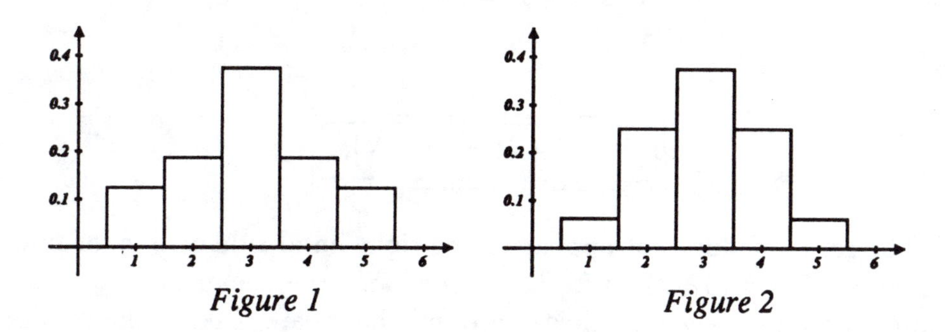

Figure 1 *Figure 2*

Sample Problem 1. *Compute the variance of the random variable X whose probability function and values are given in the following table:*

$$\begin{array}{rcllll} x_i & : & -2 & 2 & 3 & 4 \\ P(X = x_i) & : & 0.2 & 0.3 & 0.4 & 0.1 \end{array}$$

SOLUTION: As in the previous discussion about variance, the first thing to do is calculate the expected value:

$$\mu = E[X] = -2 \cdot 0.2 + 2 \cdot 0.3 + 3 \cdot 0.4 + 4 \cdot 0.1 = 1.8$$

A table showing the probability function of X will help calculate $E[X^2]$. There is a new wrinkle here because both $(-2)^2$ and 2^2 equal 4. Therefore, $P(X^2 = 4) = P(X = -2) + P(X = 2) = 0.2 + 0.3 = 0.5$ the probability function of X^2 is given by:

$$\begin{array}{rclll} x_i^2 & : & 4 & 9 & 16 \\ P(X^2 = x_i^2) & : & 0.5 & 0.4 & 0.1 \end{array}$$

The calculations

$$E[X^2] = 4 \cdot 0.5 + 9 \cdot 0.4 + 16 \cdot 0.1 = 7.2$$

and

$$V[X] = 7.2 - (1.8)^2 = 7.2 - 3.24 = 3.96$$

provide the final numerical answer. □

The analogue of the formula

$$E[aX + b] = aE[X] + b$$

for the variance is

$$\boxed{V[aX + b] = a^2 V[X].}$$

The reason b does not appear in the right hand side of this formula is that adding b to aX shifts the values and the mean but does not change the way the values cluster around the mean. Taking the square root of both sides of this equation produces

$$\boxed{\sigma[aX + b] = |a|\sigma[X]}$$

because $\sqrt{a^2} = |a|$.

For the geometric and hypergeometric random variables there are special formulas for computing the variance which are more efficient than direct calculation. The formula for the variance of a geometric random variable is simply

$$\sigma^2 = \frac{q}{p^2}.$$

The formula for the variance of a hypergeometric random variable is more complicated and can be found in most standard statistics books.

Sample Problem 2. *In the game of Parcheesi a player can only start moving a playing piece around the board by rolling doubles with a pair of dice. Let X be the number of rolls it takes a player to get his or her first piece started. Find the mean, variance, and standard deviation of X.*

SOLUTION: This is a geometric random variable with $p = 1/6$ and $q = 5/6$. From the previous section

$$\mu = \frac{1}{p} = \frac{1}{1/6} = 6.$$

The variance is

$$\sigma^2 = \frac{q}{p^2} = \frac{1/6}{(5/6)^2} = \frac{5}{6} \cdot \frac{36}{1} = 30$$

and

$$\sigma = \sqrt{30} = 5.477$$

is the standard deviation. □

The notation $P(X = a)$ can easily be modified to denote the probability of the value of X being at most or at least a. This is done by replacing $=$ with $\leq$ or $\geq$. Specifically, $P(X \leq a)$ denotes the probability that the value of X is less than or equal to a and $P(X \geq a)$ denotes the probability that the value of X is greater than or equal to a. Similarly, the probability that the value of X is greater than a and less than b is written $P(a < X < b)$.

The degree to which the standard deviation forces the values of any random variable to cluster near its mean is the point of Chebyshev's inequality. Let X be a random variable with expected value or mean μ and standard deviation σ. *Chebyshev's inequality* asserts that the probability of X differing from μ by less than k standard deviations is at least $1 - (1/k^2)$. Using the notation from the previous paragraph this can be written simply as

$$\boxed{P(\mu - k\sigma < X < \mu + k\sigma) \geq 1 - \frac{1}{k^2}.}$$

Although k is often a positive integer, this is not necessary; however the inequality is meaningless for $k \leq 1$.

To illustrate the role of the standard deviation σ in Chebyshev's inequality we will assume $\mu = 100$ and $k = 2$ and write down the inequality for three different standard deviations, namely,$\sigma = 1, 5,$ and 10. When $k = 2$,

$$1 - (1/k^2) = 1 - (1/4) = 3/4 = 0.75,$$

σ	$\mu - 2\sigma$	$\mu + 2\sigma$	CHEBYSHEV
1	98	102	$P(98 < X < 102) \geq 0.75$
5	90	110	$P(90 < X < 110) \geq 0.75$
10	80	120	$P(80 < X < 120) \geq 0.75$
15	70	130	$P(70 < X < 130) \geq 0.75$

In each of these three inequalities we are at least 75 per cent certain that the value of X is in a specific interval, but the interval grows larger as σ grows larger.

Sample Problem 3. *By collecting data over a period of time a bank has determined that it can expect 250 customers between 6:00 and 8:00 pm on Friday with a standard deviation of 15. How certain can the bank be that the number of customers on a Friday evening will be more than 200 but fewer than 300?*

SOLUTION: Let X be the number of customers entering the bank on a Friday evening. The problem is to estimate $P(200 < X < 300)$. Since the mean $\mu = 250$ is the mid-point of the interval from 200 to 300, Chebyshev's inequality can be used. This requires two steps - first determine k and second calculate the value

$$1 - \frac{1}{k^2}.$$

To determine k set

$$300 = \mu + k\sigma = 250 + k \cdot 15$$

and solve for k

$$50 = k \cdot 15$$

$$\frac{50}{15} = \frac{10}{3} = k.$$

(Alternatively we could have solved the equation $200 = 250 - k \cdot 15$ and obtained the same answer.)

Now that k has been determined the right hand side of Chebyshev's inequality can be calculated as

$$1 - \frac{1}{k^2} = 1 - \frac{1}{\left(\frac{10}{3}\right)^2} = 1 - \frac{9}{100} = \frac{91}{100} = 0.91$$

and putting the pieces together in Chebyshev's inequality

$$P(200 < X < 300) \geq 0.91.$$

Notice that we have not determined $P(200 < X < 300)$ precisely but have only given a lower estimate on its size. The answer is the bank can be at least 91 per cent certain that the number of customers on a Friday evening will be between 200 and 300, certainly adequate information on which to make staffing decisions. □

We conclude this chapter with one final but important observation about discrete random variables. All the essential information about a discrete random variable is contained in its probability function. Once the probability function of a random variable X on a sample space S has been determined, the sample space can be discarded. For example, to apply Chebyshev's inequality one needs to know μ and σ, and both μ and σ can be determined from the probability function. In fact, if X and Y have identical probability functions, then their means are equal, their variances are equal, and their standard deviations are equal.

Exercises

1. The following table gives the probability function for a random variable X:

x_i	:	-2	0	1	3	4	5
$P(X = x_i)$	:	1/20	3/20	6/20	5/20	4/20	1/20.

   Compute the expectation and variance of X.

2. The values and probability function of a random variable X are given by the following table:

x_i	:	-2	-1	1	2
$P(X = x_i)$	:	0.1	0.2.	0.4	0.3.

   Calculate the expected value, variance, and standard deviation of X.

3. Compute the variance for the outcome of a loaded die whose probability function appears below.

x_i	:	1	2	3	4	5	6
$P(X = x_i)$	:	1/6	1/4	1/6	1/3	0	1/12.

4. Compute the variance for the random variables in Exercises 1, 3, and 6 in Section 4.1 and in Exercises 1, 2, and 3 in Section 4.2.

5. Let X be a random variable with expected value 17 and variance 4. Compute the following:

   (a) $V(X - 17)$,

   (b) $E(3X - 43)$,

   (c) $V(3X - 43)$,

   (d) $\sigma(7X + 25)$,

   (e) $\sigma(-5X + 14)$,

   (f) $V(1/3X + 8)$.

6. Let W be a random variable with $\sigma(W) = 6$. Find the variance of $2W + 11$. If $W = 3Y + 7$, what is the variance of the random variable Y?

7. The probability function of a random variable Y is given below. Compute the variance of $2Y + 29$.

y_i	:	-1	0	1	2	3
$P(Y = y_i)$	:	1/16	1/8	5/8	1/8	1/16.

8. The following table gives the probability function of a random variable X:

x_i	:	15	35	40	45	60	65
$P(X = x_i)$	:	0.1	0.1	0.2	0.3	0.2	0.1.

   (a) Find the probability function of $Y = (1/5)X - 9$.

   (b) Calculate $E(Y)$ and $V(Y)$.

   (c) Now use $E(Y)$ and $V(Y)$ to compute $E(X)$ and $V(X)$.

9. The probability function of a random variable X is given by the following table:

x_i	:	-17	-11	-8	1
$P(X = x_i)$	:	0.1	0.45	0.3	0.15.

   Let $Y = (X + 11)/3$. Compute $E(Y)$ and $V(Y)$, and then use your answers to determine $E(X)$ and $V(X)$.

10. Let X be the distance that a certain make of car will run on one gallon of gas. If $\mu = 16$ and $\sigma = 1.5$, what does Chebyshev's inequality tell you about the probability that X is between 12 and 20?

11. Let X be the number of calls a fire company responds to in a week. Suppose $\mu = 25$ and $\sigma = 3$. What does Chebyshev's inequality tell you about the probability that X is between 10 and 40? What would the answer be if the standard deviation was 6 instead of 3?

12. An insurance company determines that their expected gain on a certain policy is \$12 and the standard deviation is \$3. Use Chebyshev's inequality to get a lower estimate for the probability that their gain on this policy is between 4 and 20 dollars.

13. The number of customers purchasing gas on Saturdays at a certain gas station has a mean of 175 with standard deviation σ. Use Chebyshev's inequality to obtain a lower estimate for the probability that the number of gas customers varies between 165 and 185 for $\sigma = 3, \sigma = 6$ and $\sigma = 9$.

14. Let W be a random variable with $\mu = 5$ and $\sigma = 1$. Find a lower estimate for $P(28 < -2W+50 < 52)$. (Hint: Let $X = -2W+50$.)

15. Compute the standard deviation of the random variable X with probability function:

$$\begin{array}{rcccccc} x_i & : & 5 & 15 & 20 & 30 \\ P(X = x_i) & : & 1/10 & 3/10 & 4/10 & 2/10. \end{array}$$

    (Hint: Work with $Y = (1/5)X - 3$.)

4.4 Review Problems

1. Let X be a random variable and let $Y = -2X + 3$. If $E(Y) = 11$ and $\sigma(X) = 7$, find $E(X)$ and $\sigma(Y)$.

2. If someone were to give us \$2.00 each time we roll a 6 with a fair die, how much should we give him when we roll a 1, 2, 3, 4 or 5 to make the game fair? Compute the variance of your gain.

3. The number of minutes you wait in line at the bank has a mean of 15 with a standard deviation of 3. Use Chebyshev's inequality to obtain a lower bound for the probability that you will wait between 5 and 25 minutes this afternoon.

4. Let Y be a random variable and let $X = 2Y + 21$. If $E(X) = 35$ and $V(X) = 12$, find $E(Y)$ and $V(Y)$.

5. An urn contains 3 white balls and 2 red balls. Three balls are randomly selected at the same time. Let X indicate the number of red balls selected. Find the probability function of X.

6. You roll a pair of dice. If the outcome is double 4's, 5's or 6's your opponent pays you \$6.00. For double 1's, 2's or 3's he pays you \$2.00. When you lose (i.e. fail to throw doubles), you pay him \$1.00. Whom does the game favor? What should you pay instead of \$1.00 when you lose in order to make the game fair?

7. The following table gives the values and probabilities of a random variable X:

$$\begin{array}{rcccccc} x_i & : & -2 & 0 & 1 & 3 & 4 \\ P(X = x_i) & : & 1/12 & 2/12 & 2/12 & 4/12 & 3/12. \end{array}$$

   Compute the expectation and variance.

8. At a state fair there is a gambling game in which a player pays to draw one capsule from an urn containing fifty capsules. The player keeps any money contained in the capsule. The urn always contains one capsule with a \$50 bill in it, two with \$20 bills in them, three with \$10 bills in them, and four with \$5 bills in them.

How much should the operator charge so that his expected gain each time the game is played is $1.00?

9. Let A, B and C be independent events in a sample space S with $P(A) = 4/7, P(B) = 2/3$ and $P(C) = 3/5$. The usual Venn diagram for the three events A, B and C will show which points in S belong to all three of these events, which points belong to exactly two of these events, etc.. Let X be the random variable that assigns to each point in S the exact number of these events to which the point belongs. For example, X assigns the value 3 to points in $A \cap B \cap C$ and the value 1 to points in $A' \cap B' \cap C'$. Compute the mean and variance of X.

10. A pair of fair dice, one red and one green, are rolled. Let X be the outcome of the green die minus the outcome of the red die. Find the probability function of X.

11. At a charity fair you pay a fee to toss a fair coin three times. If no heads appear you win $5.00, if exactly one heads appear you win $3.00, and you don't win anything if more than one heads appears. The organizers of the fair have set the fee to play so that their expected gain on each game is $.25. What is the fee?

12. Let X and Y be random variables with $X = (1/2)Y - 3$. Given that $E(Y) = 16$ and $E(X^2) = 34$, compute the following:

    (a) $E(X)$

    (b) $V(X)$

    (c) $\sigma(Y)$.

13. A special deck consists of 6 red and 14 white cards. The red cards are numbered from 1 to 6. If you draw a red card you win the number of dollars equal to the number on the card. (For example, if you draw the red card numbered 3 you win $3.00.) If you draw a white card you win nothing. If the fee charged to draw a card is $1.50, what is your expected gain on one play of this game?

14. Use Chebyshev's inequality to find a lower bound for $P(40 < X < 60)$ if $E(X) = 50$ and $V(X) = 4$. Same problem with $V(X) = 9$.

15. A county park wants to limit the number of people in the park to 1,200. To do this they decide to limit the number of cars entering the park so that the expected number of people in the park is 1,200. Based on a random sample of cars they believe the probability function of X, the number of people per car is:

x_i	1	2	3	4	5	6	7 or more
$P(X = x_i)$	0.1	0.3	0.25	0.2	0.1	0.05	0.

How many cars should they admit to the park?

16. Let X be a random variable with $\mu = 40$ and $\sigma = 2$.

    (a) What is $E(2X - 7)$?

    (b) What is $V(5X + 79)$?

    (c) The probability that X is between 28 and 52 is at least what?

17. Using a fair coin and a die you play the following game with your room-mate: First you roll the die. When you get a 3 or a 6 on the die, you toss the coin twice; otherwise you toss the coin only once. If you get at least one head your room-mate gives you \$5.00. What is the probability that you will win \$5.00? How much should you pay your room-mate when you do not get any heads, to make this game fair? Given that you won \$5.00 what is the probability that you got a 3 or 6 on the die?

18. The following table gives the probability function of a random variable X:

x_i	-2	0	1	3	4
$P(X = x_i)$	1/6	1/6	1/6	2/6	1/6.

Find the expectation (mean) and the variance of $4X + 9$.

19. In a casino game, an urn contains 15 colored balls - 4 blue, 5 red, and 6 white. After paying a fee, the player simultaneously draws 2 balls at random. If they are both blue, he wins 30 dollars; if they are both red, he wins 21 dollars; and if they are both white, he wins 9 dollars.

(a) Compute the probabilities of winning \$30, \$21, and \$9 respectively.

(b) What fee should the casino charge so that its expected gain is one dollar?

20. A small ice cream shop gives the following distribution of sales of ice cream cones over 30 consecutive days:

Number Sold	Number of Days this Quantity Sold
100	2
150	9
200	6
250	7
300	6

Let X denote the daily sales of ice cream cones with the thirty days considered as equally likely simple events. Compute the expected value and standard deviation of X. (Hint: Work with $Y = (1/50)X - 3$.)

21. As part of a promotional scheme the manufacturer of a new breakfast food offers one prize of \$100,000 to someone willing to try the new product (distributed without charge) and send in his name on the label. The winner will be drawn at random from all the entries. Is it worth the 20 cents postage to send in one's name if a total of 2,000,000 people send in their names?

22. A pair of fair dice is rolled. Let X be the sum of the outcomes of the individual dice. Compute the variance of X. (Hint: Work with $X - 7$.)

23. While in a barroom another customer asks you to play the following game: Both of you select a card at random from a standard deck. The card drawn by the first person is not replaced before the second card is drawn. If the colors of the cards selected are the same, you win \$1; if they are different you lose \$1. Is this a fair game?

24. Let X be a random variable with mean μ and standard deviation σ. Find k such that $P(|X - \mu| < k\sigma) \geq 45/49$.

25. A carnival game costs $.25 to throw 3 darts at balloons fastened to a board, and the player is given a prize worth $1.00 for breaking two balloons and a prize worth $10.00 for breaking three. The observed probability that a player will break a balloon on the throw of one dart is 0.2. What is the carnival's expectd gain on each play of this game?

26. An urn contains 5 green balls and 5 yellow balls. You draw two balls simultaneously. If the two balls are of the same color, you win $5. If the two balls are of different colors, you lose $4. Do you expect to win or lose money, and how much do you expect to win or lose each time you play?

Chapter 5

Continuous Random Variables

There are many random variables for which it is not feasible to make a list of their values. This is particularly true of physical measurements. For example, the weight of a randomly selected tomato is possibly any real number between 0 and 5 pounds. (According to the "Guinness Book of World Records" the largest known weight of a tomato is 4 lb. 4 oz..) This is somewhat of an idealization; when measurements are actually made we will not get arbitrary real numbers. If the scale used to weigh the tomato is only accurate to a tenth of a pound, then only $5 \cdot 10 = 50$ values would be possible. Since there is always the possibility of using a more accurate scale, it is preferable to think of the weight of the tomato as a real number between 0 and 5 and to decide later on the degree of accuracy required in the collection of data. These remarks apply to all the so called continuous random variables.

There is a bonus in the preceding argument. Many discrete random variables can be assumed to be continuous with the effect of introducing a slight error in the calculation of probabilities. For example, let X be the number of people taking a driving test at a certain state motor vehicle office during one business day. Clearly X is discrete and has a finite number of integer values. By examining the values of this random variable over a year or more, we might see a pattern similar to that of a known and tabulated continuous random variable Y. Even though the idea that half a person could take a driving test is ridiculous, for ease of calculating probabilities it would pay us to accept this idea as an idealization of the actual situation and to assume that the discrete random variable X equals the continuous random variable Y. Some other discrete random variables which might be replaced by a continuous one are the scores on a test given to a large number of people and the amount of a random sale at a fast food restaurant.

Although the emphasis in this chapter is on normal variables, there are many other types of continuous random variables which are both well understood and useful. The first section of this chapter contains a primitive introduction to the idea of a probability density for a continuous random variable, while the second section is devoted entirely to normal random variables.

5.1 Probability Densities

Let X be a random variable on a sample space S. The random variable X is *continuous* if its values are an entire interval of real numbers and $P(X = x) = 0$ for all values x. The quantities which are usually thought of as continuous random variables are time, weight, length, distance, temperature, etc..

Because $P(X = x)$ is always zero the probability function is a useless concept for a continuous random variable and other ideas must be developed to work with continuous random variables. However, by using area to represent probability the histogram paradigm can be modified for this new context.

In the study of continuous random variables $P(X < b)$ - the probability that the value of X is less than the real number b - plays a central role. In these notes $P(X < b)$ will be treated geometrically using area. (In advanced courses the calculus is used extensively to study continuous random variables.) The key concept is the probability density of a continuous random variable. The *probability density* of a continuous random variable X is a curve which lies in a plane with a coordinate system and which satisfies the following conditions:

- No portion of the curve lies below the horizontal axis
- The area of the region between the curve and the horizontal axis and to the left of the vertical line through b equals $P(X < b)$.

Figure 1 shows a probability density in which the area of the shaded region is $P(X < b)$. The area of the shaded region is usually referred to as the area beneath the curve to the left of b. The height of a probability density above the horizontal axis at any point can be thought of as the rate at which probabilities are accumulating at this point. In particular, values near high points on a probability density occur more frequently than those near low points.

The height of a probability density may be zero on portions of the axis; the restriction is only that it does not go below the axis. In fact, the examples in the rest of this section all have probability densities that have zero height for long stretches. (Look ahead at Figures 4 and 5.) When a probability density has height zero on an interval, it simply means the values in the interval are not possible for that random variable.

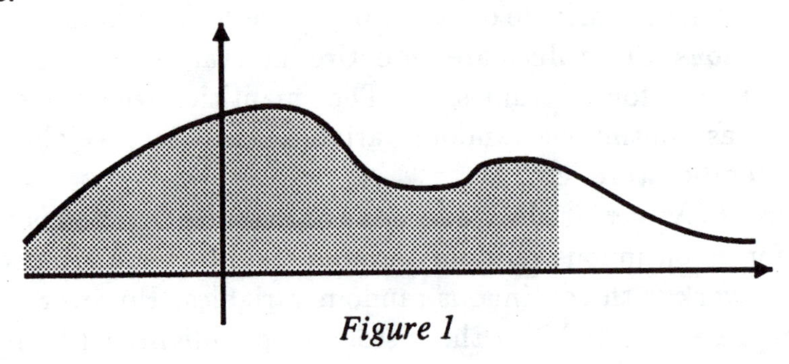

Figure 1

Because the conditions $X < b$ and $X = b$ determine disjoint events

$$P(X \leq b) = P(X < b) + P(X = b)$$

holds for any random variable X. When X is continuous $P(X = b) = 0$ and

$$P(X \leq b) = P(X < b)$$

Similarly

$$P(X \geq a) = P(X > a)$$

and

$$P(a \leq X \leq b) = P(a < X < b)$$

for continuous random variables. Thus for continuous random variables we can work entirely with strict inequalities replacing $\leq$ with $<$ at will.

To continue the discussion of the probability densities let X be a continuous random variable. For a very large real number b, the event that $X < b$ is almost a sure thing and $P(X < b)$ is nearly equal to one. When b is very large, $P(X < b)$ is also approximately equal to the total area under the probability density curve for X. It follows intuitively that the total area beneath the probability density must be 1.

Now consider the area beneath the probability density for X to the right of a (the shaded region in Figure 2). Because the total area beneath the density is 1, this shaded area must be 1 minus the area beneath the density to the left of a, which is precisely $1 - P(X < a)$. Since $1 - P(X < a)$ is the probability of the complement of the event determined by $X < a$, the area beneath the probability density to the right of a equals $P(X \geq a) = P(X > a)$.

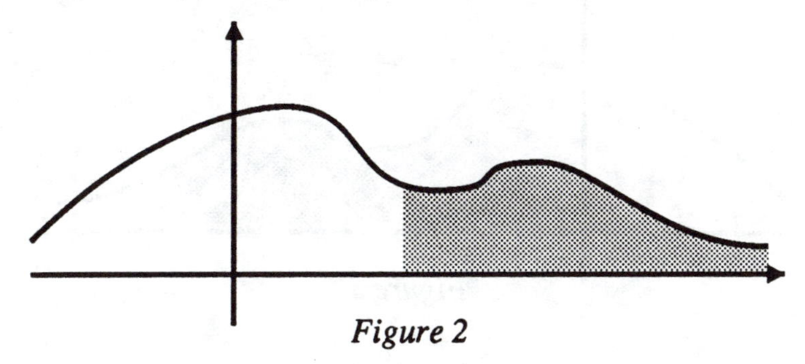

Figure 2

The conditions $X \leq a$ and $a < X < b$ determine disjoint events whose union is the event determined by the condition $X < b$. Consequently

$$P(X < b) = P(X \leq a) + P(a < X < b)$$

or

$$P(X < b) = P(X < a) + P(a < X < b)$$

because X is a continuous random variable. Subtracting $P(X < a)$ from both sides produces

$$P(X < b) - P(X < a) = P(a < X < b)$$

If we subtract the area beneath the probability density of X to the left of a from the area beneath the density to the left of b, what remains is the area beneath the probability density between a and b. Therefore, $P(a < X < b)$ equals the area beneath the probability density between

a and b as shown in Figure 3.

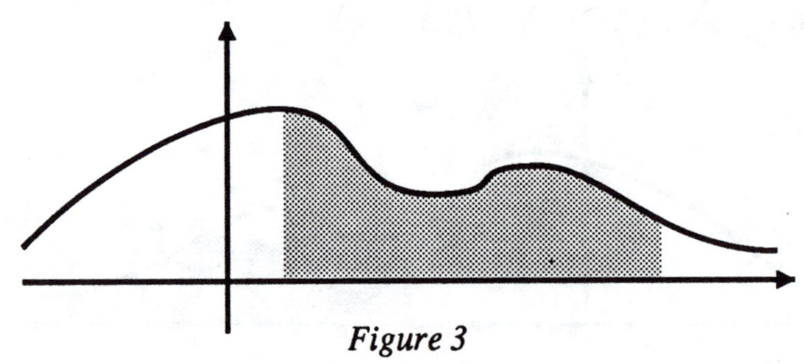

Figure 3

To summarize, for the probability density of any continuous random variable X the following are true:

- $P(X < a) =$ the area beneath the density to the left of a
- $P(X > a) =$ the area beneath the density to the right of a
- $P(a < X < b) =$ the area beneath the density between a and b.

EXAMPLE 1. Consider the triangular shaped density in Figure 4 of a continuous random variable X. The base of the large triangle has length 2 and the its height is 1. Since the area of a triangle is one half the base times the height, the area of this triangle is $(1/2) \cdot 2 \cdot 1 = 1$ as required of all probability densities. The area of the smaller shaded triangle is the probability that X is less than one or symbolically $P(X < 1)$. The base of the shaded triangle is 1 and its height is $1/2$. (The triangles are similar and the base of the smaller one is one half that of the larger. Hence the height of the smaller is one half that of the larger.) By calculating the area of the shaded triangle

$$P(X < 1) = \frac{1}{2} \cdot 1 \cdot \frac{1}{2} = \frac{1}{4}$$

and

$$P(X > 1) = 1 - \frac{1}{4} = \frac{3}{4}$$

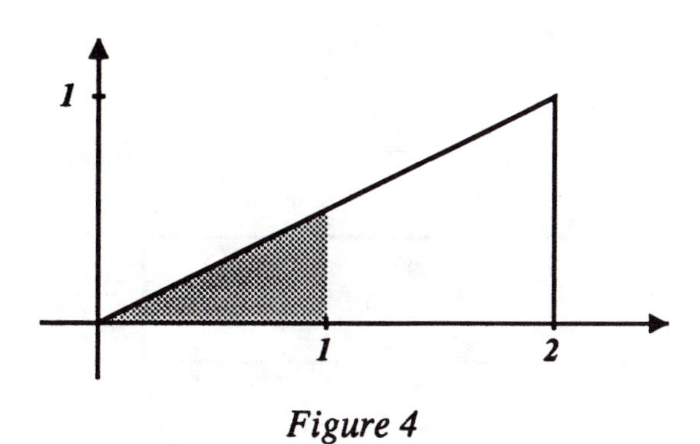

Figure 4

The simplest but still important class of continuous random variables is the uniformly distributed ones. The continuous random variable X is *uniformly distributed* over the interval from a to b if the height of its probability density is a positive constant between a and b and zero elsewhere. The probability density in Figure 4 is not uniformly distributed because its height is increasing between 1 and 2.

For any interval of real numbers, $a \leq x \leq b$, we can construct a uniform probability density by drawing a straight line above this interval at height $1/(b-a)$. By drawing the line at this height the total area under the curve is

$$\text{base times height } = (b-a) \cdot \frac{1}{(b-a)} = 1$$

which is an essential feature of probability densities.

EXAMPLE 2. Selecting a real number at random from the interval $0 \leq x \leq 1$ is an example of a continuous random variable with a uniform probability density. It is also the continuous random variable that is simulated by the random number generator on many calculators.

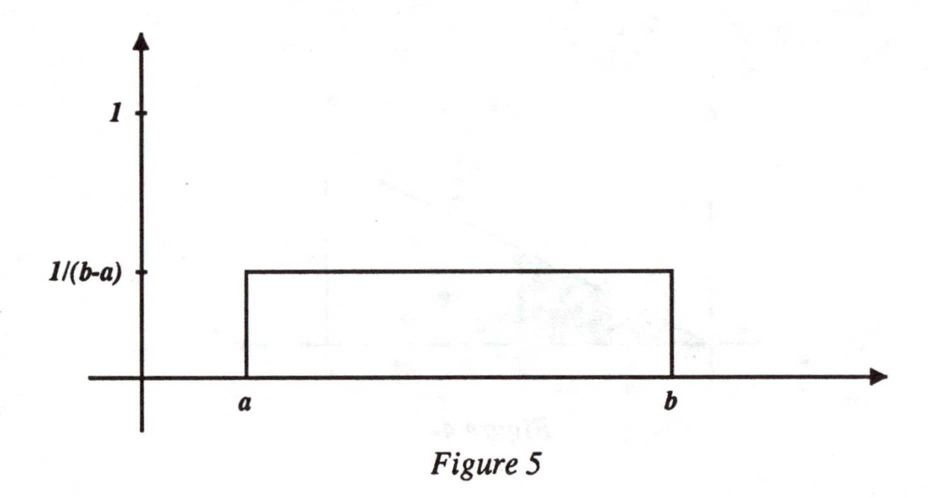

Figure 5

Figure 5 shows a typical uniform probability density for a continuous random variable X which is uniformly distributed from a to b. Consider a small interval of length L lying between a and b. Then the probability that the value of X is in this interval is the area of the rectangle above the interval and below the density. Since the height of the density is $1/(b-a)$, the probability that the value of X is in this interval is

$$L \cdot \frac{1}{b-a} = \frac{L}{b-a}.$$

The point of the discussion is that this probability depends only on the length of the interval not its location. In other words, we have equal probabilities that the value of X is in either of two intervals of equal length lying between a and b. It is this property of uniform distributions that makes them useful.

Sample Problem 1. *Let X be uniformly distributed on the interval from 1 to 3. Compute the following probabilities:*

(a) $P(X < 0.75)$,

(b) $P(X < 4)$,

(c) $P(X < 2.5)$,

(d) $P(X > 2.25)$,

(e) $P(1.25 < X < 1.75)$.

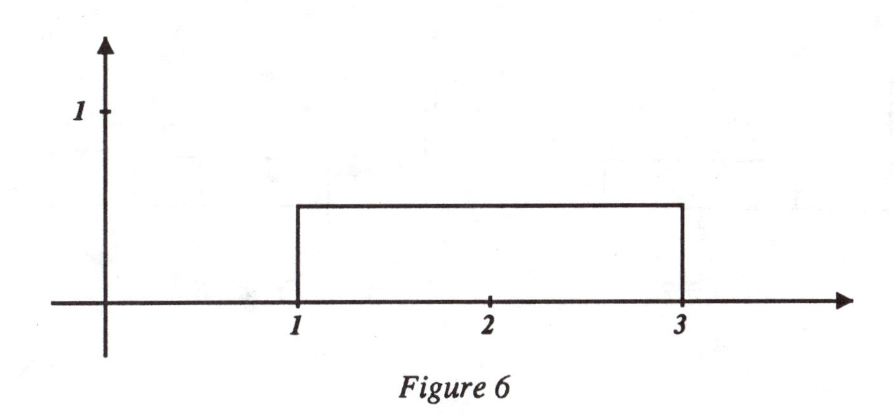

Figure 6

SOLUTION: The first step is to draw the probability density for X. Since the length of the interval from 1 to 3 is 2, the height of this uniform density must be 1/2. It is shown in Figure 6.

(a) Since all the area under the curve is to the right of 1 and $0.75 < 1$, it is impossible for X to be less than 0.75 and $P(X < 0.75)$ must be 0.

(b) Similarly $P(X < 4) = 1$ because all the area is to the left of 3 and $3 < 4$.

(c) The shaded area in Figure 7 equals $P(X < 2.5)$ and

$$P(X < 2.5) = 1.5 \cdot 0.5 = 0.75.$$

(d) We know that $P(X > 2.25)$ is the area beneath the density to the right of 2.25, which is a rectangle with base $3 - 2.25 = 0.75$ and height 0.5. Clearly

$$P(X > 2.25) = 0.75 \cdot 0.5 = 0.375.$$

(e) In like manner, $P(1.25 < X < 1.75)$ is the area under the density between 1.25 and 1.75, which is also a rectangle, and hence

$$P(1.25 < X < 1.75) = (1.75 - 1.25) \cdot 0.5 = 0.5 \cdot 0.5 = 0.25.$$

The rectangles for both parts (d) and (e) are pictured in Figure 8. □

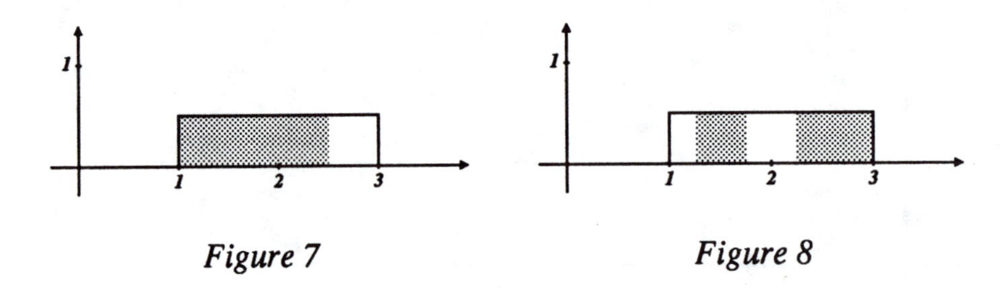

Figure 7 *Figure 8*

Sample Problem 2. *An office water cooler holds a 5 gallon jug of water. What is the probability that between 2 and 4 gallons of water are in this jug at noon?*

SOLUTION: Let X be the number of gallons of water in the jug at noon. Then X is continuous random variable whose range of values is the interval $0 \leq x \leq 5$. The problem is to compute $P(2 < X < 4)$ Since there is no reason to believe that the value of X is more likely to be in one interval of a certain length lying between 0 and 5 than in another interval of the same length, it is reasonable to assume that X has a uniform probability density. The probability density for X is shown in Figure 9 and the region whose area equals $P(2 < X < 4)$ is shaded. From this figure it is clear that

$$P(2 < X < 4) = 2 \cdot 0.2 = 0.4.$$

This is a typical situation in which a uniform probability density provides a good model. Another one is how long you wait for a subway when the trains are running every ten minutes because your arrival time on the platform is more than likely uniformly distributed over the ten minutes between trains. □

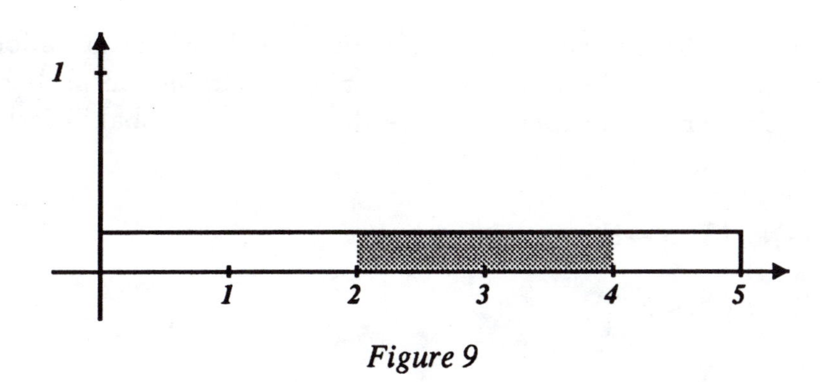

Figure 9

Continuous random variables also have expectations and variances. Although $\mu = E(X)$ and $\sigma^2 = V(X)$ are defined using calculus when X is a continuous random variable, they still have roughly the same interpretations and obey the same formulas as in the discrete case.

For the reader who knows some calculus

$$\mu = E(X) = \int_{-\infty}^{\infty} tf(t)dt$$

and

$$\sigma^2 = \int_{-\infty}^{\infty} (t-\mu)^2 f(t)dt$$

where the graph of the function $f(t)$ is the probability density of the continuous random variable X.

Exercises

1. Suppose X is a continuous random variable which is uniformly distributed on the interval $-2 \leq x \leq 2$. Graph the probability density for X and then compute the following probabilities:

   (a) $P(X > -1)$,

   (b) $P(X < 1/2)$,

   (c) $P(X > 3)$,

   (d) $P(1/4 < X < 3/4)$.

2. Suppose X is a continuous random variable which is uniformly distributed on the interval $-1 \leq x \leq 2$. Graph the probability density for X and then compute the following probabilities:

   (a) $P(X < 1)$,

   (b) $P(X > 3/2)$,

   (c) $P(-1/2 < X < 1/4)$,

   (d) $P(X > -2)$.

3. The probability density of a continuous random variable X is shown below in Figure 10. Determine the value of c and then compute both $P(X > 3/2)$ and $P(X < 3/2)$. (Hint: the area of a triangle is $(1/2)bh$.)

4. The probability density of a continuous random variable is shown below in Figure 11. Determine the value of c and then compute

$P(X < -1/2)$ and $P(-1/2 < X < 1/2)$.

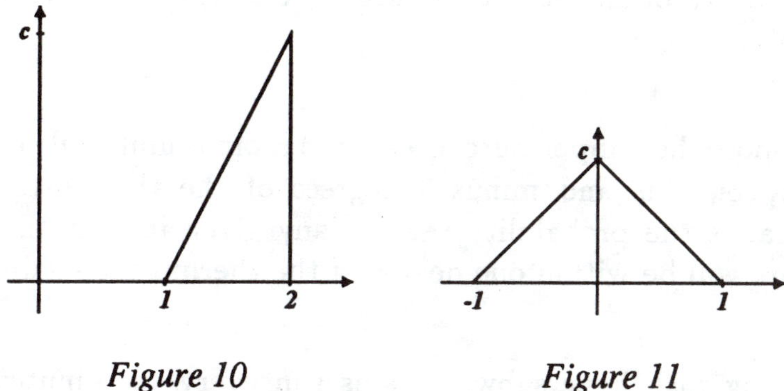

Figure 10 *Figure 11*

5. What is the probability that at noon there is less than one and a half gallons of water in the water cooler from Sample Problem 2 in the text? Same question for more than one gallon?

6. When you stop at an intersection for a red light, the length of time you must wait until the light turns green is a continuous uniformly distributed random variable. Suppose the light in your direction can remain red for at most $1.5 = 3/2$ minutes. What is the probability that you will have to wait for more then a minute for a green light? Less than 3/4 of a minute? Between a quarter of a minute and half a minute?

7. Let X be a continuous random variable which is uniformly distributed on the interval $1 \leq x \leq 5$.

   (a) Find c so that $P(X < c) = 3/4$.

   (b) Find c so that $P(X < c) = 2/5$.

   (c) Find c so that $P(X > c) = 3/8$.

8. Let X be uniformly distributed on $-6 \leq x \leq 2$.

   (a) Find c so that $P(X > c) = 1/2$.

   (b) Find c so that $P(-5 < X < c) = 1/4$.

   (c) Find c so that $P(X < c) = 3/10$.

9. Let X be the depth of the tire tread on the left rear wheel of a car selected at random. Assume X is uniformly distributed between 0 and 5/8 of an inch. Compute $P(X > 1/2)$ and $P(1/8 < X < 3/8)$.

10. Suppose the temperature in a heated room is uniformly distributed between plus and minus 3 degrees of the thermostat setting. What is the probability that at any given instant the temperature will be within one degree of the thermostat setting?

11. During rush hour subway trains run every 10 minutes. If you arrive on the platform at a random time, what is the probability that you will have to wait more than 3 minutes for the next train?

12. The formulas for the mean and variance of a continous random variable X which is uniformly distributed on $a \leq x \leq b$ are

$$\mu = (b + a)/2 \text{ and } \sigma^2 = (b - a)^2/12.$$

Compute μ and σ^2 for the random variables in Exercises 1, 2, 6, and 11.

13. The probability density of a continuous random variable X is shown below. Calculate the following probabilities:

(a) $P(X < 1)$,

(b) $P(X < 2)$,

(c) $P(X < 1/2)$,

(d) $P(X > 1/2)$,

(e) $P(.5 < X < 1.5)$.

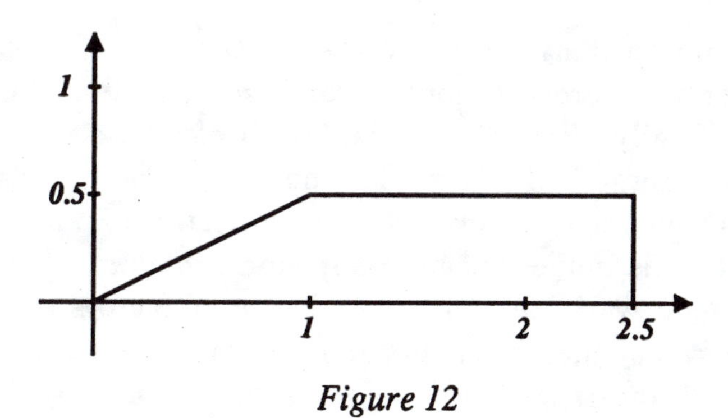

Figure 12

14. The remnant table in a fabric store is piled high with pieces of fabric of lengths from 1 to 6 feet. It is reasonable to assume the lengths of these remnants is uniformly distributed. If you select a remnant at random, what is the probability that its length will be at least one yard?

15. When you order a quart of hand dipped vanilla ice cream in a dairy store, it is reasonable to assume the amount of ice cream in the store's open 5 gallon container is uniformly distributed. What is the probability that the sales person will have to open a new 5 gallon container to fill your order?

5.2 Normal Random Variables

Probably the most useful and certainly the most famous continuous random variables are the normal random variables. They consist of an infinite family of random variables all of whose probability densities have the same bell shape. The normal random variables are also ubiquitous which is the source of their importance. Virtually any measurement that is influenced by many small random effects produces a normal random variable. For example, many factors contribute to errors in laboratory measurements and the error - the difference between the measured and true value - is known to be normally distributed. In fact, the standard normal is also know as the error function. Because of their prevalence and well understood probability densities, normal random variables provide an effective statistical model for the study of a wide range of problems.

Each normal random variable is determined by its mean μ and standard σ. To work with this infinite family of random variables we single out the normal random variable with $\mu = 0$ and $\sigma = 1$ and use it as a reference or standard. The probability density of this standard normal variable will be described geometrically.

The *standard normal random variable*, which will always be denoted by Z, has a probability density with the following geometric properties:

- It has a single maximum at $z = 0$.
- It is symmetric about the vertical axis.
- It decreases as you move away from $z = 0$ in either direction.
- It is bell shaped.

The probability density for Z is shown in Figure 1. Moreover, Z also has the following properties:

$$\begin{aligned} E(Z) &= 0 \\ V(Z) &= 1 \\ \sigma(Z) &= 1 \end{aligned}$$

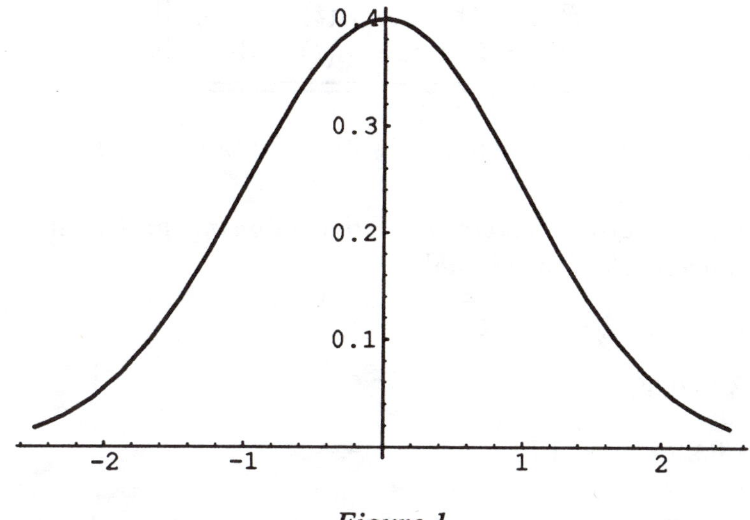

Figure 1

By symmetry the area under the standard normal density curve to the left of the vertical axis must equal the area under the curve to the right of the vertical axis. Since the total area under a probability density is always 1, both these areas must equal 1/2 and

$$P(Z < 0) = \frac{1}{2} = 0.5.$$

This is the only numerical value for Z which can be easily computed. To calculate other probabilities for the standard normal random variable a table will be used.

For convenience let $A(t)$ denote the area beneath the standard normal density to the left of t, so that

$$P(Z < t) = A(t).$$

Numerical values for $A(t)$, can be found in Table I in the Appendix. This table, in conjunction with the ideas from the previous section, will be used to determine probabilities for normal random variables.

The three basic principles for probability densities from the previous section imply that

$$\boxed{\begin{array}{l} P(Z < t) = A(t) \\ P(Z > t) = 1 - A(t) \\ P(s < Z < t) = A(t) - A(s). \end{array}}$$

The next Sample Problem shows how Table I is used.

Sample Problem 1. *Determine the following probabilities for the standard normal random variable:*

(a) $P(Z < 1.27)$,

(b) $P(Z > 0.32)$,

(c) $P(-0.95 < Z < 1.5)$.

SOLUTION: The answer to each part will depend on using Table I to determine a specific numerical value of $A(t)$.

(a) First $P(Z < 1.27) = A(1.27)$ which can be found in Table I. To do so find 1.2 in the extreme left hand column of numbers, move to the right in this row until you are in the column headed by 0.07, and read off the four digit number 0.8980, that is

$$P(Z < 1.27) = A(1.27) = 0.8980.$$

(b) This part requires a preliminary step before the table can be used, namely,

$$P(Z > 0.32) = 1 - A(0.32).$$

Now from the table $A(0.32) = 0.6255$ and

$$P(Z > 0.32) = 1 - .6255 = 0.3745.$$

(c) For the last part use the formula $P(s < Z < t) = A(t) - A(s)$ and then the table. Specifically

$$\begin{aligned} P(-0.95 < Z < 1.5) &= A(1.5) - A(-0.95) \\ &= 0.9332 - 0.1711 \\ &= 0.7621. \end{aligned}$$

It is often helpful to make a little sketch of the area corresponding to the probability in question. This is especially true when the solution to the problem is not obvious to you. The sketches which might accompany the solution to the above example are shown in Figure 2. □

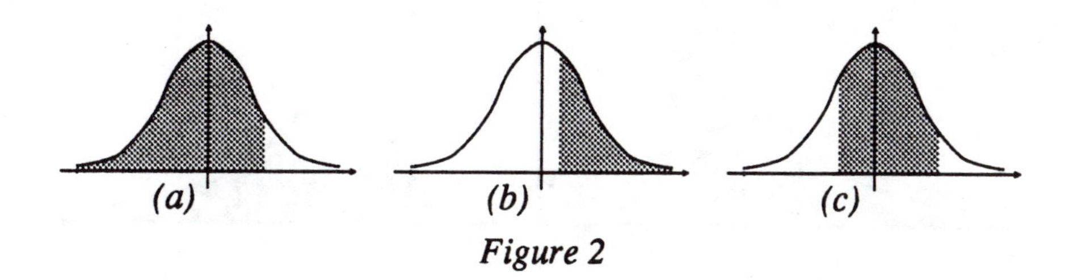

Figure 2

From the symmetry of the standard normal density curve, it follows that the areas under the standard normal density to the left of $-t$ and to the right of t are equal. (See Figure 3.) In other words,

$$P(Z < -t) = P(Z > t).$$

Upon replacing $P(Z > t)$ by $1 - P(Z < t)$, it becomes

$$P(Z < -t) = 1 - P(Z < t)$$

or

$$A(-t) = 1 - A(t).$$

A useful consequence of this observation is the following:

$$\begin{aligned} P(-t < Z < t) &= A(t) - A(-t) = A(t) - [1 - A(t)] \\ &= A(t) - 1 + A(t) = 2A(t) - 1. \end{aligned}$$

Thus

$$\boxed{P(-t < Z < t) = 2A(t) - 1.}$$

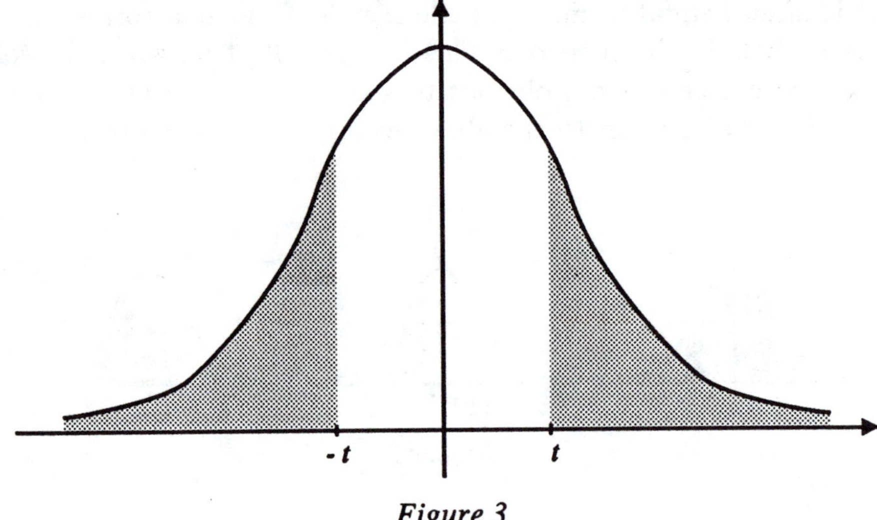

Figure 3

The use of Table I can be reversed. More concretely, given the probability that Z is less than t, Table I cam be used to find the value of t.

EXAMPLE 1. If it is known that $P(Z < t) = 0.8023$, we can try to determine t. Since $P(Z < t) = 0.8023$ is the same as $A(t) = 0.8023$, we need only find 0.8023 in the center of Table I and read off the t value from the edges of this table to see that $t = 0.85$. The next Sample Problem illustrates several variations of this theme.

Sample Problem 2. *Find the value of t such that*

(a) $P(Z > t) = 0.6026$,

(b) $P(0 < Z < t) = 0.2517$,

(c) $P(-t < Z < t) = 0.8324$.

SOLUTION: Each part requires some preliminary work to put the equation in the form $A(t)$ equals a specific number so that Table I can be used as in the previous Example.

(a) First use the fact that $P(Z > t) = 1 - P(Z < t)$ to rewrite

$P(Z > t) = 0.6026$ as

$$1 - P(Z < t) = 0.6026.$$

Then because $P(Z < t) = A(t)$, we have

$$1 - A(t) = 0.6026$$

or

$$1 - 0.6026 = 0.3974 = A(t)$$

and from the table $t = -0.26$

(b) The equation $P(0 < Z < t) = 0.2517$ is the same as

$$A(t) - A(0) = 0.2517$$

or

$$A(t) - 0.5 = 0.2517.$$

By adding 0.5 to both sides we get $A(t) = 0.7517$ and from the table $t = 0.68$.

(c) From the formula $P(-t < Z < t) = 2A(t) - 1$ it follows that

$$2A(t) - 1 = 0.8324.$$

Using simple algebra to solve for $A(t)$, we obtain

$$A(t) = 0.9162$$

and $t = 1.38$ from the table.

These questions were all written so that the numbers appeared in the table. When this not the case, just find the closest value in the table. □

Thus far our development of normal random variables has been limited to the standard normal with mean zero and standard deviation one. To construct a *normal random variable with mean μ and standard deviation σ, $\sigma > 0$*, let $X = \sigma Z + \mu$. Then

$$\begin{aligned} E(X) &= E(\sigma Z + \mu) \\ &= \sigma E(Z) + \mu = \sigma \cdot 0 + \mu = \mu \end{aligned}$$

because $E(Z) = 0$ and

$$\begin{aligned} V(X) &= V(\sigma Z + \mu) \\ &= \sigma^2 V(Z) \\ &= \sigma^2 \end{aligned}$$

because $V(Z) = 1$. Of course $V(X) = \sigma^2$ implies

$$\sigma(X) = \sqrt{V(X)} = \sqrt{\sigma^2} = \sigma.$$

The statement that X is a *normal random variable with mean μ and standard deviation σ* means precisely that

$$X = \sigma Z + \mu.$$

The probability densities of the standard normal random variable Z and the normal random variable $X = \sigma Z + \mu$ are very similar. The maximum or high point of the density of X occurs at μ instead of 0. In other words, the addition of μ just translates the bell shape to the right or left according as μ is positive or negative. The effect of σ is on the height of the bell. For large σ the bell is low and wide while for small σ the bell is high and narrow. (See Figure 4.)

Notice that for small σ most of the area under the bell shaped normal density is close to the mean $\mu = 0$ and thus the probability that the value of X is close to μ is relatively large. Consequently the smaller σ is, the more likely it is that the value of X is near μ. This agrees with our intuitive understanding of the standard deviation for a discrete random variable.

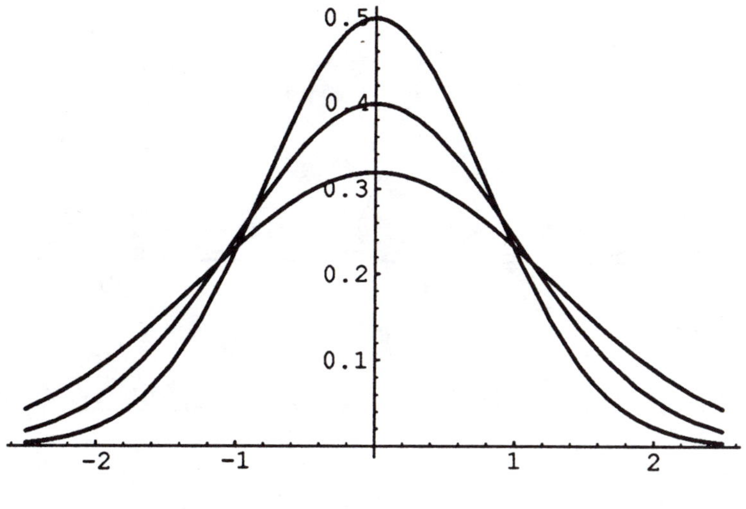

Figure 4

Probabilities for the normal random variable $X = \sigma Z + \mu$ can be calculated by referring to the standard normal random variable Z and using Table I. The basic procedure is illustrated by the following calculation:

$$\begin{aligned} P(X < a) &= P(\sigma Z + \mu < a) \\ &= P(\sigma Z < a - \mu) \\ &= P\left(Z < \frac{a - \mu}{\sigma}\right) \\ &= A\left(\frac{a - \mu}{\sigma}\right). \end{aligned}$$

All computations of probabilities for $X = \sigma Z + \mu$ will be done in this manner - subtract the mean μ , divide by the standard deviation σ, and use the table.

Sample Problem 3. *Let X be a normal random variable with $\mu = 75$ and $\sigma = 5$. Determine the following probabilities:*

(a) $P(X < 78)$

(b) $P(X > 86)$

(c) $P(65 < X < 75)$

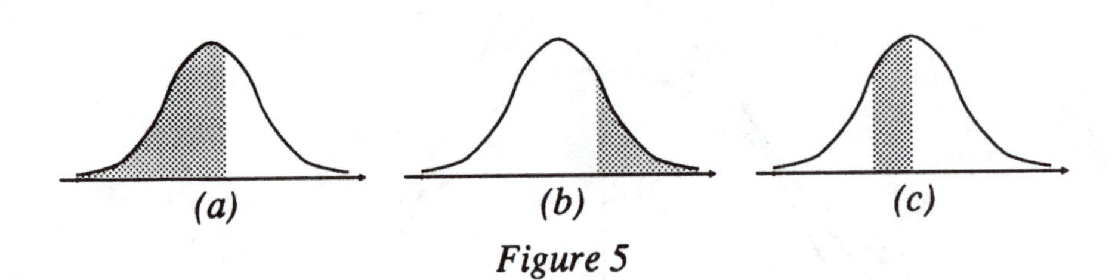

Figure 5

SOLUTION: Figure 5 gives a sketch of each area being determined.

(a) As above

$$\begin{aligned} P(X < 78) &= P(5Z + 75 < 78) \\ &= P(5Z < 78 - 75) \\ &= P(Z < 3/5) \\ &= A(0.6) = 0.7257. \end{aligned}$$

With some practice it is easy to eliminate some of the intermediate steps.

(b) Although the main idea is the same as in part (a), some additional small steps appear in the computation

$$\begin{aligned} P(X > 86) &= 1 - P(X < 86) \\ &= 1 - P\left(Z < \frac{86 - 75}{5}\right) \\ &= 1 - A\left(\frac{11}{5}\right) \\ &= 1 - A(2.2) \\ &= 1 - 0.9861 = 0.0139. \end{aligned}$$

We chose to us the complement first and then subtract the mean and divide by the standard deviation. We could have just as well subtracted the mean and divided by the standard deviation first to show that $P(X > 86) = P(Z > 2.2)$ and then proceeded as before in Sample Problem 1 to calculate $P(Z > 2.2)$.

(c) Here we work with both inequalities at the same time.

$$\begin{aligned} P(68 < X < 75) &= P(65 < 5Z + 75 < 75) \\ &= P\left(\frac{65-75}{5} < Z < \frac{75-75}{5}\right) \\ &= P(-2 < Z < 0) \\ &= A(0) - A(-2) \\ &= 0.5 - 0.0228 \\ &= 0.4772. \end{aligned}$$

After subtracting the mean and dividing by the standard deviation, this solution is almost identical to the solution of Sample Problem 1. □

The methods of solution used in the previous example can be expressed as general formulas. For normal random variable X with mean μ and standard deviation σ the following are true:

$$\begin{aligned} P(X < b) &= A\left(\frac{b-\mu}{\sigma}\right) \\ P(X > a) &= 1 - A\left(\frac{a-\mu}{\sigma}\right) \\ P(a < X < b) &= A\left(\frac{b-\mu}{\sigma}\right) - A\left(\frac{a-\mu}{\sigma}\right). \end{aligned}$$

When the left hand side of any one of these equations is known, then one of the terms a, b, μ, or σ can be an unknown. The next Sample Problem contains some questions of this type.

Sample Problem 4. *Let X be a normal random variable with mean μ and standard deviation σ .*

(a) If $\mu = 30$ and $\sigma = 2$, find k such that $P(X < k) = 0.8810$

(b) If $P(X > 50) = 0.2514$ and $\sigma = 10$, find μ.

(c) If $P(25 < X < 30) = 0.3944$ and $\mu = 25$, find σ.

SOLUTION: The solution parallels Sample Problem 2 except there is some additional preliminary work before Table I can be used and a little algebra after using it.

(a) The equation $P(X < k) = 0.8810$ is the same as

$$A\left(\frac{k-30}{2}\right) = 0.8810.$$

Looking at the central part of Table I, we find that $A(1.18) = 0.8810$. Consequently 1.18 must equal $(k - 30)/2$, i.e.

$$1.18 = \frac{k-30}{2}$$

which can be solved for k. Thus

$$2 \cdot 1.18 + 30 = k$$

or $k = 32.36$. Parts (b) and (c) will be solved in a similar manner.

(b) Because $P(X > 50) = 1 - P(X < 50)$, the given information about $P(X > 50)$ can be written

$$1 - P(X < 50) = 0.2514$$

or

$$1 - A\left(\frac{50-\mu}{10}\right) = 0.2514.$$

Before Table I can be used, $A([50 - \mu]/10)$ must be determined. Simple algebra shows that

$$1 - 0.2514 = A\left(\frac{50 - \mu}{10}\right)$$

or

$$0.7486 = A\left(\frac{50 - \mu}{10}\right).$$

Now from the table we obtain

$$0.67 = \frac{50 - \mu}{10}$$

from which it follows that

$$\mu = 50 - 6.7 = 43.3.$$

(c) From $P(25 < X < 30) = 0.3944$ and $\mu = 25$ it follows that

$$A\left(\frac{30 - 25}{\sigma}\right) - \left(\frac{25 - 25}{\sigma}\right) = 0.3944$$

or

$$A(5/\sigma) - A(0) = 0.3944.$$

Then because $A(0) = 0.5$

$$A(5/\sigma) = 0.3944 + 0.5 = 0.8944,$$

and with the help of Table I

$$\frac{5}{\sigma} = 1.25.$$

Finally,

$$\sigma = \frac{5}{1.25} = 4.$$

This completes the solution. □

Almost any numerical measurement taken from a human or animal population as well as many physical measurements are approximately normally distributed. This means that a graph with the observed measurements on the horizontal axis and their frequency of occurrence on the vertical axis has a bell shape like a normal probability density. Although we can not expect a perfect normal density, in most cases the difference will not be significant; this is especially true when a large amount of data is used. Consequently, if X denotes the approximately normally distributed quantity being measured, it can be assumed that

$$X = \sigma Z + \mu$$

without causing undue error. Of course, to make effective use of this approximation both σ and μ must be determined. They are usually obtained from the data as sample means and sample standard deviations. The importance of the normal random variables is simply that in many different situations one can safely assume $X = \sigma Z + \mu$ and effectively calculate relevant probabilities and use them to make decisions.

Sample Problem 5. *The weight of roasting chickens sent to retail stores by a chicken processing plant is approximately normally distributed with mean μ and standard deviation $\sigma = 0.3$.*

(a) *Given that $\mu = 4.5$ what is the probability that a roaster from this plant will weigh at least 4 pounds?*

(b) *The plant can control the mean without changing σ by varying the age of the chickens it processes. For what mean will 85% of their roasters weigh less than 6 pounds?*

SOLUTION: Let X be the weight of a roasting chicken. Because X is approximately normally distributed, it is reasonable to set $X = 0.3Z + \mu$.

(a) The problem is to compute $P(X > 4)$ which is done as follows:

$$\begin{aligned} P(X > 4) &= 1 - P(X < 4) \\ &= 1 - A\left(\frac{4 - 4.5}{0.3}\right) = 1 - A\left(\frac{-0.5}{0.3}\right) \\ &= 1 - A(-1.666\ldots) = 1 - A(-1.67) \\ &= A(1.67) = 0.9525. \end{aligned}$$

(b) The requirement that eighty-five per cent of the roasters weigh less than six pounds can be written as

$$P(X < 6) = 0.85.$$

Since X is a normal random variable, this is the same as

$$A\left(\frac{6-\mu}{0.3}\right) = 0.85.$$

The entry in Table I closest to 0.85 is $0.8508 = A(1.04)$, so we set

$$\frac{6-\mu}{0.3} = 1.04$$

and solve for μ to get

$$\mu = 5.688.$$

When determining $A(t)$ in most situations, it suffices to round off t to 2 decimal places. Conversely, to find t from $A(t)$ it is usually sufficient to use the nearest entry to the given $A(t)$. □

Let X be a normal random variable with mean μ and standard deviation σ , i.e. $X = \sigma Z + \mu$. Let k be a positive number and consider

$$P(\mu - k\sigma < X < \mu + k\sigma) = P(\mu - k\sigma < \sigma Z + \mu < \mu + k\sigma)$$

which is just the probability that X differs from its mean by less than k standard deviations. Subtracting μ and dividing by σ as usual produces

$$P(\mu - k\sigma < X < \mu + k\sigma) = P(-k < Z < k) = 2A(k) - 1.$$

Since neither μ nor σ appears in the expression $2A(k) - 1$, the probability that X differs from its mean by less than k standard deviations, $P(\mu - k\sigma < X < \mu + k\sigma)$, is the same for all normal random variables. If we use the table to compute $2A(k) - 1$ for $k = 1, 2,$ and 3, we get $0.6826, 0.9544,$ and 0.9974 respectively. Hence, the value of any normal random variable is within 1, 2, or 3 standard deviations of its mean 68%, 95% or 99% of the time respectively.

Exercises

1. Determine the following probabilities for the standard normal random variable Z:

   (a) $P(Z < 1.95)$,
   (b) $P(Z > 0.57)$,
   (c) $P(Z < 0.09)$,
   (d) $P(0.79 < Z < 2.35)$.

2. Determine the following probabilities for the standard normal random variable Z:

   (a) $P(Z > 2.44)$,
   (b) $P(0 < Z < 1.23)$,
   (c) $P(1 < Z < 2)$,
   (d) $P(Z > 0.35)$.

3. Compute the following probabilities for the standard normal random variable Z:

   (a) $P(Z < -1.73)$,
   (b) $P(Z > -3.07)$,
   (c) $P(-0.37 < Z < 0.73)$,
   (d) $P(-1.83 < Z < 1.83)$.

4. Let Z be the standard normal random variable. Find the following:

   (a) $P(Z < -1.06)$,
   (b) $P(-1.34 < Z < 1.34)$,
   (c) $P(Z > -1.75)$,
   (d) $P(-0.52 < Z < -0.07)$,
   (e) $P(Z > 1.27)$,
   (f) $P(-1.5 < Z < 0.5)$.

5. As usual let Z be the standard normal random variable. Find t such that:

   (a) $P(Z < t) = 0.9515$,
   (b) $P(-t < Z < t) = 0.1034$,
   (c) $P(Z > t) = 0.3936$,
   (d) $P(Z < t) = 0.1038$.

6. In each of the following determine t from the probability for the standard normal random variable Z:

   (a) $P(-t < Z < t) = 0.8324$,
   (b) $P(0 < Z < t) = 0.4871$,
   (c) $P(Z > t) = 0.0336$,
   (d) $P(Z > t) = 0.7764$.

7. Let X be a normal random variable with $\mu = 100$ and $\sigma = 10$. Calculate the following:

   (a) $P(X < 117)$,
   (b) $P(X > 103)$,
   (c) $P(85 < X < 115)$,
   (d) $P(X < 88)$,
   (e) $P(81 < X < 114)$,
   (f) $P(X > 82)$.

8. Let X be a normal random variable with $\mu = 60$ and $\sigma = 4$. Calculate the following probabilities:

   (a) $P(X < 63)$,
   (b) $P(X > 61)$,
   (c) $P(55 < X < 65)$,
   (d) $P(X < 53)$,
   (e) $P(62 < X < 71)$.

9. Let X be a normal random variable with mean 42 and standard deviation 3.

   (a) Find k so that $P(X < k) = .9306$
   (b) Find k so that $P(X < k) = .2611$
   (c) Find k so that $P(X > k) = .0060$
   (d) Find k so that $P(X > k) = .6255$
   e) Find k so that $P(42 < X < k) = .3997$

10. Let X be a normal random variable with $\mu = 18$ and $\sigma = 5$. Compute the following probabilities:

    (a) $P(15 < X < 25)$,
    (b) $P(X > 20)$,
    (c) $P(X < 10)$,
    (d) $P(X > 0)$.

11. The weights of new born babies are approximately normally distributed with $\mu = 7.5$ pounds and $\sigma = 1.25$. What is the probability that a new born baby will weigh more than 9 pounds? Between 6.5 and 8.5 pounds?

12. The IRS finds that the lengths of time of calls for tax information are approximately normally distributed with $\mu = 8$ and $\sigma = .5$. What is the probability that the length of such calls will be between 8 and 9 minutes? Less than $63/4$ minutes?

13. The annual snowfall in a northern city is normally distributed with a mean of 45 inches and a standard deviation of 10 inches.

    (a) What is the probability that in a particular year the snowfall is between 38 and 47 inches?
    (b) What is the probability the snowfall is greater than 40 inches?

14. A certain study claims that the weight of twenty-five year old females who are 5 feet 8 inches tall is normally distributed with a mean of 130 pounds and a standard deviation of 10 pounds.

(a) What is the probability that such a woman would weigh between 120 and 145 pounds?

(b) What is the weight above which only 5% of such women would weigh?

15. Let X be a normal random variable with expected value 27 and variance 25. In each of the following solve for k:

(a) $P(X > k) = 0.0594$,

(b) $P(X < k) = 0.1515$,

(c) $P(22 < X < k) = 0.1394$.

16. The results of a mathematics test given to a large number of students are approximately normally distributed with mean 67 and standard deviation 15. Below what grade are 90% of the scores? If the faculty decides that scores in the upper 20% but not upper 10% constitute B work, what grades should they tell the students are B's?

17. Let X be a normal random variable with unknown mean and standard deviation 2.

(a) If $P(X < 95) = .9147$, what is μ?

(b) If $P(X > 88) = .4168$, what is μ?

18. The number of hours a 100 watt bulb will operate before burning out is approximately normally distributed with standard deviation $\sigma = 50$.

(a) If the mean $\mu = 900$, what is the probability that a light bulb will burn out in less than 966 hours of use?

(b) By varying the quality of the filament used in the bulb, a company can change the mean. What mean should a company use if it wants the probability that a bulb burns out after 800 hours of use to be .03?

19. The amount of fat in a pound of hamburger from a grocery store is approximately normally distributed with a mean of 5 ounces per pound and a standard deviation of 1.6 ounces. What is the probability that a pound of hamburger will contain more than 7 ounces of fat? The company can control the mean by how they trim the meat used to make hamburger. What mean should they choose to be 80% certain that a pound will contain at most 6 ounces of fat?

20. The standard deviation of a normal random variable X is 20. Given that $P(X < 200) = .9821$, find μ.

21. Let X be a normal random variable with mean $\mu = 21.8$. If $P(X < 35) = .9505$, what is the standard deviation?

22. A weekly poker player observes over the years that his weekly winnings have been approximately normally distributed with mean $13.35 and that 75% of the time he has won at least $10.00. What is the standard deviation of his winnings?

23. The length of time before a concrete block of sidewalk poured by the F. B. Night Company begins to crumble is approximately normally distributed with $\mu = 15$ years and $\sigma = 4$ years.

    (a) What is the probability that a block selected at random from those poured by this company does not start to crumble for at least 18 years?

    (b) What is the probability that one of its blocks will begin crumbling between 10 and 15 years after it is poured?

    (c) The company can control the mean without changing the standard deviation by varying the amount of sand in the concrete. What mean should it use so that the probability that one of its blocks begins to crumble within 7 years is .063?

24. The demand for seats on the 12:00 shuttle to New York is approximately normally distributed with $\mu = 80$ and $\sigma = 8$. What is the probability that more than 70 people will want a seat on this flight on a given day?

25. A study shows that the number of calories per day in the diets of 20 year old males in the U.S. is normally distributed with a mean of 3300 calories and a standard deviation of 320 calories. However, medical science recommends only 2900 calories per day for this population. What per cent of 20 year old males exceed this dietary recommendation?

26. Let X be a normal random variable with mean μ and standard deviation σ.

    (a) If $\mu = 130$ and $\sigma = 7$, find k so that $P(130 - 7k < X < 130 + 7k) = .75$

    (b) If $\sigma = 7$, find k so that $P(\mu - 7k < X < \mu + 7k) = .75$.

    (c) Find k so that $P(\mu - \sigma k < X < \mu + \sigma k) = .75$.

5.3 Review Problems

1. The number of miles per year that cars in the United States are driven is approximately normally distributed with $\mu = 12,000$ miles and $\sigma = 1,500$ miles. What is the probability that a car is driven

   (a) less than $14,000$ miles a year,

   (b) less than $7,500$ miles a year,

   (c) between $11,000$ and $13,000$ miles a year,

2. The recorded temperature in a freezer is uniformly distributed between $-8°$C and $0°$C. What is the probability that at any given time the temperature will be above $5°$C? What is the probability that at any given time the temperature will be below $-2°$C?

3. Let Z be the standard normal random variable.

   (a) Determine $P(Z > 1.27)$.

   (b) Determine $P(-0.75 < Z < 0.25)$.

   (c) Determine t such that $P(Z < t) = 0.1292$.

   (d) Determine $P(Z > -0.6)$.

   (e) Determine t such that $P(-t < Z < t) = 0.7698$.

   (f) Determine t such that $P(1 < Z < t) = 0.1385$.

4. Let X be a normal random variable with a standard deviation of 20. If $P(X < 50) = 0.8531$, what is the mean of X?

5. Let X be a uniformly distributed random variable on the interval $-2 \leq x \leq 4$. Calculate

   (a) $P(X < 0)$,

   (b) $P(1/2 < X < 2)$,

   (c) $P(X > 1/3)$,

   (d) $P(-1/2 < X < 3)$,

   (e) $P(X = 1)$,

(f) $P(X > -3)$.

6. The weights of adult laboratory rats fed on a standard diet are normally distributed with a mean of 10 ounces and a standard deviation of 2 ounces.

   (a) What is the probability that a rat picked at random will weigh between 8 and 11 ounces?

   (b) By changing the diet of the rats, the mean of their weights can be changed without changing the standard deviation.For what mean weight will the probability that a rat weighs at least 12 ounces be .0968?

7. A certain size box of a common cereal is supposed to contain 18 ounces of cereal, but there is some variation in the weight of the contents caused by the packaging machinery. Assume the weight of the cereal in a box is normally distributed with mean 18 ounces and standard deviation .2 ounces. What is the probability that the contents of a box of this cereal will weigh less than 17.67 oz.? Between 17.45 oz. and 17.95 oz.?

8. Two economists agree that the weekly earnings of people working in manufacturing is normally distributed with mean \$321.00, but they disagree on the standard deviation. The first estimates that $\sigma = \$12.00$ and the second estimates that $\sigma = \$25$. Calculate the per cent of people working in manufacturing whose weekly earnings exceed \$300 for both their estimates of σ.

9. Compute the conditional probability $P(1 < X < 2 | X < 3/2)$ for X uniformly distributed on the interval $0 \leq x \leq 5$.

10. The height of male students at the University of Maryland is approximately normally distributed with mean 70 inches and standard deviation 4 inches.

    (a) What is the probability that a student chosen at random will be at most 68 inches tall?

    (b) What is the probability that a student chosen at random will be more than 63 inches tall?

(c) Use Chebyshev's inequality to obtain a lower bound for the probability that the height of a student selected at random will be between 60 and 80 inches.

d) Compute the probability that the height of a randomly selected male student is between 60 and 80 inches and compare the result with the answer to (c).

11. Let X be a normal random variable with a mean of 82. If $P(X < 75) = .1587$, what is the standard deviation of X?

12. An oceanside bathhouse supplies each individual shower with a new 3 oz. cake of soap when the old one is gone. When you enter one of these showers what is the probability of finding a cake of soap weighing less than 1/2 an ounce? What is the probability of finding a cake of soap weighing between 2 and 2.75 ounces? (Hint: first decide on a reasonable continuous random variable to use.)

13. Let X be a random variable with $\mu = 35$ and $\sigma = 4$.

    (a) Determine $P(X < 34)$.

    (b) Determine $P(32 < X < 38)$.

    (c) Determine $P(X > 30)$.

    (d) Determine x so that $P(X > x) = 0.0122$.

    (e) Determine x so that $P(x < X < 35) = 0.3106$.

    (f) Determine x so that $P(X > x) = 0.6255$.

14. Consider two brands of batteries A and B. Let X be the life in hours of a brand A battery and let Y be the life in hours of a brand B battery. Assume X and Y are normally distributed with means 100 and 110 hours and standard deviations 5 and 25 respectively.

    (a) Which brand has the greater probability of lasting more than 90 hours?

    (b) Which brand has the greater probability of lasting between 100 and 110 hours?

(c) Which brand has the greater probability of lasting more than 110 hours?

15. A manufacturer of garbage disposals guarantees their product unconditionally for 18 months. (This means that if any garbage disposal fails to operate in a satisfactory manner within 18 months of the date of purchase they will replace it without any charge to the purchaser). The length of life X of one of their garbage disposals is approximately normally distributed with a mean of 20 months and a standard deviation of 4 months.

    (a) What is the probability that the manufacturer will have to replace one of their garbage disposals?

    (b) Out of every 100 units sold about how many will they usually have to replace?

    Since replacing such a high percentage would be a financial disaster for the company, they decide to increase the expected length of life of their disposals by improving materials and production techniques. The manufacturer would like to keep the 18 month guarantee and to have the probability of replacing a unit to be less than 0.02.

    (c) What value of $\mu = E(X)$ would accomplish this goal assuming the standard deviation does not change?

16. The results of a standardized test given to a large number of high school juniors range from 0 to 200 and are approximately normally distributed with mean 90 and unknown standard deviation. If 20% of the grades lie within 6 points of the mean, what is the standard deviation?

17. Compute the conditional probability $P(X > 65 | 63 < X < 75)$ for X a normal random variable with $\mu = 63$ and $\sigma = 8$

18. A standardized exam is known to produce normally distributed scores between 0 and 100 with a standard deviation of 7. When this exam is given to a particular group of people, 89.25% of them receive a score of 64 or better. What is the mean for this group?

19. The probability density of a continuous random variable X is shown below. Calculate

    (a) $P(X < 1)$,

    (b) $P(X < 2.5)$,

    (c) $P(X > 3.5)$,

    (d) $P(2.5 < X < 3.5)$.

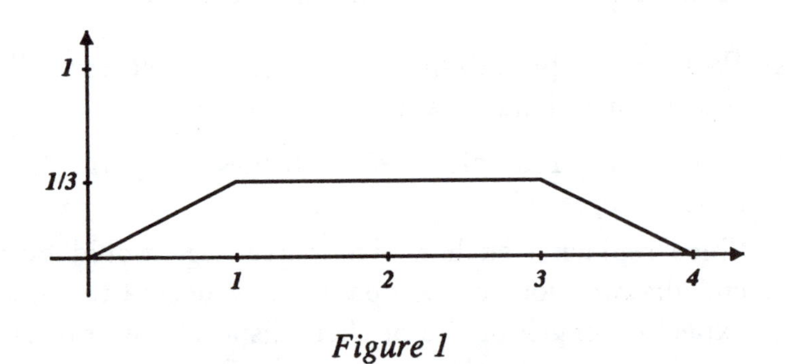

Figure 1

20. The life span of white mice is approximately normally distributed with a mean of 15 months and a standard deviation of 2.5 months.

    (a) What is the probability that a white mouse will live longer than 17 months?

    (b) What is the probability that the life span of a white mouse is between 10 and 14 months?

    (c) In a laboratory experiment using white mice 100 of them are used as a control population, i.e. they are allowed to lead normal lives. If 25 of the control mice die before they are 12 months old, would you or wouldn't you suspect an abnormal health problem in your mouse population?

21. If the scores on an exam are approximately normally distributed, and the professor decides to give C's to the students whose grades deviate from the mean by at most one half of a standard deviation, what per cent of the class will get C's?

22. The distance employees travel daily to work in the Washington area is approximately normally distributed with $\mu = 12$. If the probability that an employee selected at random in this area travels less than 8.5 miles to work is 0.242, what is the probability that he/she travels less than 13 miles to work?

23. Physical measurements like time, distance and weight are usually assumed to be normally distributed with the true value as the mean. Suppose a large number of fans time the winning horse in the Preakness and it is known that the standard deviation for such measurements is 5 seconds. If 67% of the times measured by the fans are less than 118 seconds, what is the true time of the winner?

24. One of the laboratory experiments in a science course is to measure the number of calories in a tablespoon of butter. Suppose 170 of the 200 students doing the experiment measure at least 95 calories in a tablespoon of butter. If these measurements are normally distributed with the true value as mean and a variance of 9 calories, what is the true value?

Chapter 6

Binomial Random Variables

Often in a complex situation one is only concerned with one aspect of the outcome of a random phenomenon; that is, we single out a particular event in which we are interested. Although there may be a large amount of information in the outcome of the random phenomenon, we choose to ignore everything except whether or not the event we are interested in has occurred. This produces a Bernoulli trial with success being the occurrence of our special event. Consequently, Bernoulli trials arise naturally in any random experiment. To study a particular event in this way we repeat the experiment, and hence the Bernoulli trial, independently many times and count the number of successes to create a binomial random variable.

Besides the fact that binomial random variables are very common, why should there be a whole chapter devoted to them? First they provide an excellent opportunity to bring together many of the ideas in the previous chapters. Secondly, the Law of Large Numbers for a binomial random variable sheds some light on the intimate connection between probability and relative frequency. Thirdly, the relationship between binomial and normal random variables is the classic illustration of approximating a discrete random variable by a continuous one.

A binomial random variable, which simply counts the number of times a particular unpredictable event occurs, is a rather primitive idea, but its analysis requires a wide range of ideas stretching through all five previous chapters. In addition to using the Axioms and their immediate consequences, independence is also an implicit ingredient when working with a binomial random variable because it is always assumed that the repetitions of the underlying experiment are done independently. For a small number of repetitions, combinations are essential in the calculation of the probability function of a binomial random variable, and the normal random variables save the day for a large number of repetitions. Finally the mean and standard deviation are an integral part of all calculations with normal random variables.

Intuitively we believe that the way to determine the probability of rolling a six with a loaded die is to roll the die many times and calculate the relative frequency of getting a six. Does the theory of probability confirm this belief? The Law of Large Numbers discussed in the first section of this chapter gives a positive answer. However, the relationship between relative frequency and the true probability is a very sophisticated one because there is always some degree of uncertainty. The statement that relative frequency gets closer and closer to the true probability when the experiment is repeated more and more times is not true unless we add the phrase "with higher and higher probability." The Law of Large Numbers tells us that even if we roll the die a million times there is still a small but positive probability that the relative frequency is not as close to the true probability as desired. This can be made precise. Using the Law of Large numbers we can determine how many times an experiment must be repeated so that the relative frequency approximates the true probability with both a required accuracy and degree of certainty.

Since the relative frequency is just the number of successes, a binomial random variable, divided by the number of tries, the two are closely related and binomial random variables for a large number of tries of the underlying Bernoulli trial are important. The histogram for a binomial random variable looks increasing like a normal random variable as the number of trials increases. In fact, a binomial random variable quickly becomes approximately normal when the number of tries is increased. This fact is the original version of the Central Limit Theorem discovered by de Moivre in 1773 and a fitting topic for the final section.

6.1 Basic Properties

The outcome of n tries of an independent Bernoulli trial is a list of the results, success or failure, in order of occurrence. For example, if a fair coin is tossed six times, then a possible outcome is TTTHHT which means the result is heads on only the fourth and fifth tosses. All these lists form the sample space for the experiment of repeating a Bernoulli trial n times. To compute the value of the binomial random variable one counts the number of successes in such a list. More precisely a *binomial random variable* is the number of successes in n tries of an independent Bernoulli trial. Clearly a binomial random variable is a discrete random variable with values $0, 1, 2, \ldots, n$.

What is the probability function for a binomial random variable? As usual let p denote the probability of success and $q = 1 - p$ the probability of failure. For $k = 0, 1, \ldots, n$ we need a formula for the probability function $P(X = k)$. Since the value of X is k when there are exactly k successes, $P(X = k)$ is the same as the probability of exactly k successes in n tries of a Bernoulli trial which equals

$$\binom{n}{k} \cdot p^k \cdot q^{n-k}$$

The above formula first appears at the end of Chapter 3 and is based on the following reasoning: By independence $p^k \cdot q^{n-k}$ is the probability of success in k specified trials and failure in the rest. Because there are $\binom{n}{k}$ ways to specify where the successes are to occur, the probability of exactly k successes must be $\binom{n}{k}$ times $p^k \cdot q^{n-k}$. Since this is the same as $P(X = k)$, we have established that

$$\boxed{P(X = k) = \binom{n}{k} \cdot p^k \cdot q^{n-k}.}$$

Notice that the right hand side of this formula is one of the terms that appears when the Binomial Theorem is applied to $(p + q)^n$ which explains the name binomial random variable.

Sample Problem 1. *A fair die is rolled six times. Let X be the* *number of times the outcome is a multiple of 3. Compute the probability function of X and calculate $P(X \leq 2)$.*

SOLUTION: Let success on the roll of the die be a multiple of 3, i.e. a 3 or a 6. This is certainly an independent Bernoulli trial with $p = 1/3$ and $q = 2/3$. Then X is the number of successes in six rolls of the dice which is a binomial random variable with $n = 6$. The values of X are $0, 1, 2, 3, 4, 5,$ and 6. Using the above formula the probability function of X is computed as follows:

$$\begin{aligned}
P(X=0) &= \binom{6}{0}\cdot\left(\frac{1}{3}\right)^0\cdot\left(\frac{2}{3}\right)^6 = 1\cdot\frac{2^6}{3^6} = 0.0878 \\
P(X=1) &= \binom{6}{1}\cdot\left(\frac{1}{3}\right)^1\cdot\left(\frac{2}{3}\right)^5 = 6\cdot\frac{2^5}{3^6} = 0.2634 \\
P(X=2) &= \binom{6}{2}\cdot\left(\frac{1}{3}\right)^2\cdot\left(\frac{2}{3}\right)^4 = 15\cdot\frac{2^4}{3^6} = 0.3292 \\
P(X=3) &= \binom{6}{3}\cdot\left(\frac{1}{3}\right)^3\cdot\left(\frac{2}{3}\right)^3 = 20\cdot\frac{2^3}{3^6} = 0.2195 \\
P(X=4) &= \binom{6}{4}\cdot\left(\frac{1}{3}\right)^4\cdot\left(\frac{2}{3}\right)^2 = 15\cdot\frac{2^2}{3^6} = 0.0823 \\
P(X=5) &= \binom{6}{5}\cdot\left(\frac{1}{3}\right)^5\cdot\left(\frac{2}{3}\right)^1 = 6\cdot\frac{2^1}{3^6} = 0.0165 \\
P(X=6) &= \binom{6}{6}\cdot\left(\frac{1}{3}\right)^6\cdot\left(\frac{2}{3}\right)^0 = 1\cdot\frac{1}{3^6} = 0.0014
\end{aligned}$$

When $X \leq 2$, the value of X must be $0, 1,$ or 2 and

$$\begin{aligned}
P(X \leq 2) &= P(X=0) + P(X=1) + P(X=2) \\
&= 0.088 + 0.263 + 0.329 = 0.680.
\end{aligned}$$

Notice that once the probability function has been calculated all other probabilities such as $P(X > 4)$ and $P(1 \leq X \leq 5)$ can be computed by adding the appropriate values from the probability function. □

For discrete random variables it is common to tabulate $P(X \leq k)$ instead of the probability function $P(X = k)$. It is easy to calculate one from the other. Working with $P(X \leq k)$ provides a strong analogy with normal random variables and tables for $P(Z < a)$.

EXAMPLE 1. By successively adding the terms $P(X = k)$ for $k = 0, 1, 2, 3, 4, 5, 6$ in the previous Sample Problem we obtain $P(X \leq k)$. The following table shows both $P(X = k)$ and $P(X \leq k)$ for any binomial random variable with $n = 6$ and $p = 1/3$:

k	$P(X = k)$	$P(X \leq k)$
0	0.0878	0.0878
1	0.2634	0.3512
2	0.3292	0.6804
3	0.2195	0.8999
4	0.0823	0.9822
5	0.0165	0.9986
6	0.0014	1.0000

EXAMPLE 2. A histogram can always be constructed from the probability function of a discrete random variable. Figure 1 is the histogram of the binomial random variable ($n = 6$ and $p = 1/3$) in Sample Problem 1. It was constructed using the solution to Sample Problem 1 and is a typical histogram of a binomial random variable.

Sample Problem 2. *Let X be a binomial random variable with $n = 10$ and $p = 0.4$.* (A common quick way of saying this is , "X is $b(10, 0.4)$".) *Find the following probabilities:*

(a) $P(X \leq 2)$,

(b) $P(X < 4)$,

(c) $P(X > 8)$,

(d) $P(2 \leq X \leq 6)$,

(e) $P(X = 4)$.

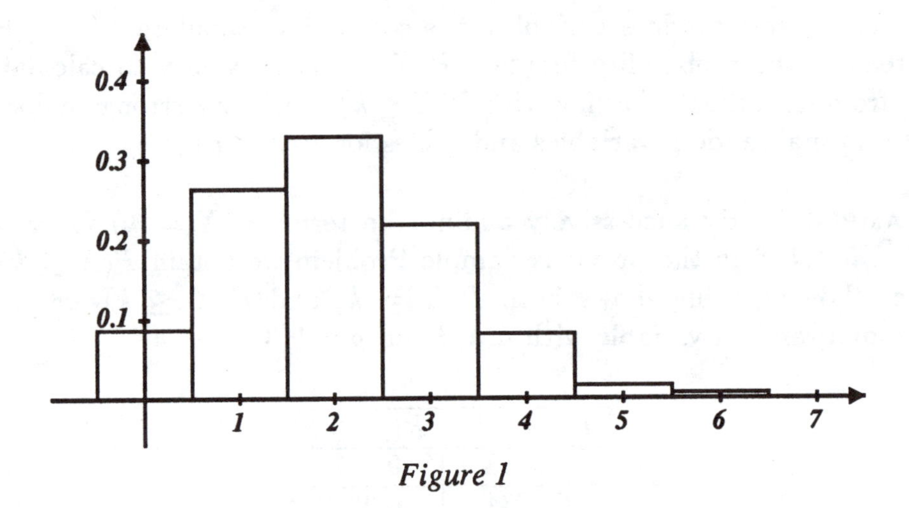

Figure 1

SOLUTION: As in the previous example one way to find $P(X \leq 2)$ is to use the formula for $P(X = k)$ and $P(X \leq 2) = P(X = 0) + P(X = 1) + P(X = 2)$. However, it is not necessary to do all this arithmetic because the results of calculations like this are tabulated. Table II in the Appendix contains $P(X \leq k)$ for $n = 2$ through 13 and $p = 0.05, 0.10, 0.15, \ldots, 0.50$. This table will be used to answer all four questions.

(a) To locate $P(X \leq 2)$ in Table II first find n equal to 10 in the far left column, then find x equal to 2 in the next column, and finally in that row read off the entry 0.1673 in the column headed by p equals 0.40 which is

$$P(X \leq 2) = 0.1673.$$

(b) For a discrete random variable $P(X < b)$ and $P(X \leq b)$ need not be the same. In this case $P(X < 4) = P(X = 0) + P(X = 1) + P(X = 2) + P(X = 3) = P(X \leq 3) = 0.3823$ from Table II as in the previous part.

(c) The complement of $X > 8$ is $X \leq 8$. Hence $P(X > 8) = 1 - P(X \leq 8) = 1 - 0.9983 = 0.0017$ where 0.9983 was obtained from Table II.

(d) From Table II $P(X \leq 6) = 0.9452$. What should we subtract from $P(X \leq 6)$ to get $P(2 \leq X \leq 6)$? Since $2 \leq X$ excludes $X = 0$ and $X = 1$, $P(X = 0) + P(X = 1) = P(X \leq 1) = 0.0860$ should be subtracted. Thus

$$\begin{aligned} P(2 \leq X \leq 6) &= P(X \leq 6) - P(X \leq 1) \\ &= 0.9452 - 0.0464 \\ &= 0.8988. \end{aligned}$$

(e) Similarly $P(X = 4) = P(X \leq 4) - P(X \leq 3) = 0.6331 - 0.3823 = 0.2508$.

Using the method used to calculate $P(X = 4)$ in part (e), it is not difficult to reconstruct the probability function for a binomial random variable with $n = 10$ and $p = 0.40$ from Table II. □

Unfortunately tables of probabilities like Table II often do not contain all the information that one might like, but usually one can find devices to extract more information. The next sample problem demonstrates how the use of Table II can be extended.

Sample Problem 3. *Suppose X is $b(6, 0.7)$. Determine*

(a) $P(X \geq 2)$,

(b) $P(X \leq 4)$.

SOLUTION: These questions can not be answered directly with Table II. However, by switching the meaning of success and failure Table III can be used. Let X' be the number of failures in 6 tries of a Bernoulli trial with $p = 0.7$. This interchanges the roles of success and failure. Since the probaility of failure for X is 0.30, the random variable X' is binomial with $n = 6$ and $p = 0.30$, which is included in Table II. The following table shows how the values of X and X' are related and enables us to easily translate questions about probabilities of X to those of X'.

X Values	X' Values
0	6
1	5
2	4
3	3
4	2
5	1
6	0

(a) It is evident from the above table that $X \geq 2$ precisely when $X' \leq 4$. Thus

$$P(X \geq 2) = P(X' \leq 4) = 0.9777.$$

(b) Observe from our table that requiring $X \leq 4$ is the same as requiring $X' \geq 2$. Now there is an added complication because Table II gives $P(X \leq k)$ and it is necessary to use the complement as follows:

$$\begin{aligned} P(X \leq 4) = P(X' \geq 2) &= 1 - P(X' \leq 1) \\ &= 1 - 0.4202 = 0.5798. \end{aligned}$$

By switching the meaning of success and failure, which really are arbitrary, the information in Table II was doubled. The preliminary step of making a table showing how the values are related when success and failure are interchanged is an effective way to avoid errors in the translation of the problem from X to X'. □

Let X be a binomial random variable. The mean or expected value of X is given by

$$\boxed{\mu = np}$$

where p is the probability of success for the underlying Bernoulli trial and n is the number of times it is repeated. This agrees with our intuition.

EXAMPLE 3. If you toss a fair coin 800 times, the natural guess for the expected number of heads is 400 which equals 1/2 or p times 800 or n.

Although the formula for the variance

$$\boxed{V[X] = \sigma^2 = npq}$$

is less intuitive, the standard deviation of X given by

$$\boxed{\sigma = \sqrt{npq}.}$$

will play an important role in the rest of the chapter.

Sample Problem 4. *The probability that a slightly defective soda machine will dispense a soda when properly operated is 3/5. During a day 150 people attempt to use the machine to purchase a soda. What is the expected number of these people who obtain a soda from this machine? With what standard deviation?*

SOLUTION: Trying to buy a soda from this machine should be viewed as a Bernoulli trial with $p = 3/5$. Moreover, it is to be repeated 150 times. Let X be the number of times the machine dispenses a soda, so X is a binomial random variable. We want to compute $E[X]$ and $\sigma[X]$

$$\begin{aligned} E[X] &= np = 150 \cdot \frac{3}{5} = 90 \\ \sigma[X] &= \sqrt{npq} = \sqrt{150 \cdot \frac{3}{5} \cdot \frac{2}{5}} \\ &= \sqrt{36} = 6. \end{aligned}$$

How might we interpret a standard deviation of 6 in this setting? Compared with 150, the total number of possible values of X, 6 is a relatively small standard deviation and we should expect the values of X far from 90 to occur with low probability. □

Notice that for X in the above example we have no means for calculating probabilities like $P(X \leq 100)$. Table II cannot be used because $n = 150$ and the table stops at $n = 13$. Carrying out the arithmetic needed to use the formula for $P(X = k)$ would be a hopeless task; nobody wants to calculate lots of terms like:

$$P(X = 100) = \binom{150}{100} \left(\frac{3}{5}\right)^{100} \left(\frac{2}{5}\right)^{50} =?$$

Developing an effective method for calculating probabilities like $P(X = 100), P(X \leq 95)$, etc., when X is a binomial random variable with n large, will be the theme of Section 6.2.

Another interesting random variable associated with the repetition of an independent Bernoulli trial is the proportion of successes. If X is the number of successes in n tries of a Bernoulli trial, then

$$Y = \frac{1}{n}X$$

is the *proportion of successes* or the relative frequency of success. We know that

$$E[X] = np \text{ and } V[X] = npq.$$

In particular, the variance will become large as n becomes large, and we cannot expect the value of X to be close to $E[X] = np$ when n is big. However, Y behaves differently.

The expected value and variance of Y can be computed using formulas from Chapter 4. First using $E[aX + b] = aE[X] + b$

$$\begin{aligned} E[Y] &= E\left[\frac{1}{n}X\right] \\ &= \frac{1}{n}E(X) = \frac{1}{n}np = p, \end{aligned}$$

or simply

$$\boxed{E[Y] = p.}$$

Next employing $V[aX + b] = a^2V[X]$

$$\begin{aligned} V[Y] &= V\left[\frac{1}{n}X\right] \\ &= \left(\frac{1}{n}\right)^2 V[X] = \frac{1}{n^2} \cdot npq = \frac{pq}{n} \end{aligned}$$

and we have

$$\boxed{V[Y] = \frac{pq}{n}}$$

Finally the standard deviation, which is the square root of the variance is given by

$$\boxed{\sigma[Y] = \sqrt{\frac{pq}{n}}.}$$

There are two important consequences of these formulas. First the variance of Y becomes small as n becomes large, which is the opposite of the behavior of $V[X]$. For example, if Y is the proportion of successes in 10, 100, or 1000 tosses of a fair coin, then the variance of Y is

$$\begin{aligned} \frac{(0.5)\cdot(0.5)}{10} = \frac{0.25}{10} &= .025 \\ \frac{(0.5)\cdot(0.5)}{100} = \frac{0.25}{100} &= .0025 \\ \frac{(0.5)\cdot(0.5)}{1000} = \frac{0.25}{1000} &= .00025 \end{aligned}$$

respectively. This means that when a fair coin is tossed 1000 times we would expect the proportion of heads to deviate very slightly from 0.5.

Secondly since we now have a random variable Y whose expected value is P, doing the experiment many times and calculating an actual value of Y should be a good procedure for estimating the value of anunknown p. Although this is an important consequence of the formula $E[X] = p$, it has limitations which Chebyshev's inequality will help us understand.

For n at most 13 Table II can also be used to compute probabilities for the proportion of success, and the next example illustrates the method for doing this.

Sample Problem 5. *Let Y be the proportion of successes in 12 tries of a Bernoulli trial with $p = 0.25$. Compute $P(0.1 < Y < 0.5)$.*

SOLUTION: Let X be the number of successes so that $Y = (1/12)X$ or $X = 12Y$. Then multiplying by 12 changes

$$0.1 < Y < 0.5$$

into the inequality

$$12 \cdot (0.1) < 12Y < 12 \cdot (0.5)$$

and then replacing $12Y$ by X produces

$$1.2 < X < 6.$$

Because the values of X are integers $1.2 < X < 6$ is equivalent to $2 \leq X \leq 5$. Therefore,

$$\begin{aligned} P(.1 < Y < .5) &= P(2 \leq X \leq 5) \\ &= 0.9456 - 0.1584 = 0.7872 \end{aligned}$$

with the probabilities coming from Table II. □

For larger n Chebyshev's inequality can be used to estimate the probable deviation between Y and its mean. Recall that Chebyshev's inequality is

$$P(\mu - k\sigma < Y < \mu + k\sigma) \geq 1 - \frac{1}{k^2}$$

where μ and σ are the mean and standard deviation of Y. Since Y is the proportion of successes, we can substitute p and $\sqrt{pq/n}$ for μ and σ to get

$$\boxed{P\left(p - k \cdot \sqrt{\frac{pq}{n}} < Y < p + k \cdot \sqrt{\frac{pq}{n}}\right) \geq 1 - \frac{1}{k^2}}$$

which is called the *Law of Large Numbers.*

What does the Law of Large Numbers tell us? Suppose a fair die was altered by filing the corners on the face containing six dots. Presumably the probability of rolling a six would no longer be 1/6. How could the probability of a six be determined? The Law of Large Numbers states that the proportion of sixes in a large number of rolls would with high probability be close to the actual probability of rolling a six with this altered die. The remainder of this section will examine in more detail how the Law of Large Numbers works in specific situations.

First we explore the role of the standard deviation, $\sqrt{pq/n}$, in

$$P\left(p - k \cdot \sqrt{\frac{pq}{n}} < Y < p + k \cdot \sqrt{\frac{pq}{n}}\right) \geq 1 - \frac{1}{k^2}.$$

The effect of a very large n in this denominator is to make pq/n and hence $k\sqrt{pq/n}$ very small. Consequently the interval from $p - k\sqrt{pq/n}$ to $p + k\sqrt{pq/n}$ is very short when n is large, and the Law of Large Numbers implies that if we repeat an independent Bernoulli trial often enough, we have a measure of the probability that the deviation of the proportion of successes from p is extremely small. In particular, the relative frequency of an event is likely to be very close to the probability of the event, but how likely?

To illustrate how this aspect of the Law of Large Numbers works with some specific numbers, suppose a fair coin is first tossed 100 times and then 10,000 times. Here $p = q = 0.5$ and $\sqrt{pq} = \sqrt{0.25} = 0.5$. Furthermore, $\sqrt{n}$ is either $\sqrt{100} = 10$ or $\sqrt{10,000} = 100$. (Beware, the numbers do not always work out so simply). Take $k = 2$ so that $1 - (1/k^2) = 1 - 1/4 = 3/4 = 0.75$. For 100 tosses of the coin

$$k\sqrt{\frac{pq}{n}} = \frac{k\sqrt{pq}}{\sqrt{n}} = \frac{2 \cdot (0.5)}{10} = 0.1$$

and from the Law of Large Numbers

$$P(0.5 - 0.1 < Y < 0.5 + 0.1) = P(0.4 < Y < 0.6) \geq 0.75.$$

Similarly for 10,000 tosses $k\sqrt{pq/n} = 2 \cdot (0.5)/100 = 0.01$ and

$$P(0.49 < Y < 0.51) \geq 0.75.$$

Thus for 100 or 10,000 tosses the proportion of heads deviates from 0.5 by at most 0.1 or 0.01 respectively with probability at least 0.75.

The number k can also be increased. What effect does that have? The following table shows how $1 - 1/k^2$ changes as k increases:

k	$1 - \frac{1}{k^2}$
2	0.7500
3	0.8889
4	0.9375
5	0.9600
6	0.9733
7	0.9796
8	0.9843
9	0.9876
10	0.9900

It is quite evident from the numbers in this table that by increasing k the probability $P(p - k\sqrt{pq/n} < Y < p + k\sqrt{pq/n})$ gets closer and closer to 1. Making both n and k large gives the full effect of the Law of Large Numbers and the probability that Y is close to p is almost equal to 1.

The Law of Large Numbers raises another question: Given k can we control the deviation of the proportion of successes Y from p by choosing n large enough? The expression $k\sqrt{pq/n}$ is both added and subtracted from p and is the amount of deviation from p allowed in the Law of Large Numbers. In the remainder of this section and the next section the deviation from p will be consistently denoted by d. In the current situation

$$d = \frac{k\sqrt{pq}}{\sqrt{n}}.$$

The deviation d can be controlled because this equation can be solved for n as follows:

$$d\sqrt{n} = k\sqrt{pq}$$

$$\sqrt{n} = \frac{k\sqrt{pq}}{d}$$

$$n = \frac{k^2pq}{d^2}.$$

Now d and k can be specified first and then n can be calculated using the above equation. Therefore, given a specified degree of accuracy and a degree of certainty, we can determine how many times to repeat the experiment to estimate p to the given specifications.

Sample Problem 6. *An independent research laboratory wants to verify the claim made by a seed company that* 80% *of its hybrid zucchini seeds will germinate. Assuming the* 80% *is correct, how many of these seeds should they plant to be* 96% *certain that the proportion which germinates will be between* 76% *and* 84%*?*

SOLUTION: The Bernoulli trial in this example is planting a zucchini seed and success means the seed germinates. Clearly $p = 0.8$ and $q = 0.2$. Let Y be the proportion of seeds germinating. The goal is to choose n so that using the Law of Large Numbers

$$P(0.76 < Y < 0.84) \geq 0.96.$$

The appropriate value of k can be read off the previous table and is $k = 5$. Next the allowed deviation d from p must be determined. From $0.76 = 0.80 - 0.04$ and $0.84 = 0.80 + 0.04$ it follows that $d = 0.04$. To complete the solution we use

$$n = \frac{k^2pq}{d^2} = \frac{25 \cdot (0.8) \cdot (0.2)}{(0.04)^2} = 2,500.$$

If they conduct this experiment with at least 2,500 seeds and only 70% germinate, they would have good reason to suspect the company's claim which is the underlying idea of hypothesis testing. □

The Law of Large Numbers justifies determining the probability of an event E by repeating the experiment and calculating the relative frequency or proportion of times the event occurs. The idea is to think of the occurrence of E as a success in a Bernoulli trial so $P(E) = p$. Then independently repeat the experiment n times and compute Y the proportion of successes which in this case is the number of times E occurred divided by n. The Law of Large Numbers guarantees that the more often we repeat the experiment the more certain we can be that Y is very close to $p = P(E)$. Note there is a catch in this procedure; we can never be 100% certain that Y is approximately equal to p. Although it is highly unlikely, it is possible to toss a fair coin 100 times, get heads 85 times, and wrongly conclude that the probability of heads is about 0.85 and that the coin is not fair.

Exercises

1. Let X be the number of successes in 4 tries of an independent Bernoulli trial with $p = 3/5$. Calculate the probability function of X and make a histogram for X.

2. Let X be the number of heads in 5 tosses of a fair coin. Calculate the probability function of X and make a histogram for X.

3. Let X be the random variable in Exercise 1. Calculate $P(X \leq k)$ for $k = 0, 1, 2, 3$ and 4.

4. Let X be the random variable in Example 1 of the text. Calculate $P(X \leq k)$ for $k = 0, 1, 2, 3,$ and 4.

5. Let X be a binomial random variable with $n = 8$ and $p = 0.4$. Use Table II to determine the following probabilities:

   (a) $P(X \leq 3)$,
   (b) $P(X < 6)$,
   (c) $P(X \geq 2)$,
   (d) $P(X = 4)$.

6. Determine the following probabilities for a random variable X whose distribution is $b(12, 1/4)$:

   (a) $P(X > 4)$,
   (b) $P(3 \leq X \leq 7)$,
   (c) $P(X = 5)$,
   (d) $P(1 < X < 6)$.

7. As in Sample Problem 3 of the text let X be $b(6, 0.7)$. Find:

   (a) $P(X \leq 2)$,
   (b) $P(X > 3)$,
   (c) $P(X = 5)$,
   (d) $P(1 \leq X \leq 4)$.

8. Let X be the number of successes in 9 tries of an independent Bernoulli trial with $p = 0.6$. Determine the following probabilities:
   (a) $P(X \leq 5)$,
   (b) $P(X = 3)$,
   (c) $P(X > 6)$,
   (d) $P(4 \leq X \leq 7)$.

9. Let X be a binomial random variable with $n = 35$ and $p = .7$. Find numerical expressions (but don't do any calculating) for the following:
   (a) $P(X = 17)$,
   (b) $P(X \geq 34)$,
   (c) $P(X > 2)$.

10. A pair of fair dice are rolled 48 times. Let X be the number of times the sum of the outcomes is 4 or 7. Also compute the mean and standard deviation of X. Use the formulas in the beginning of this section to find numerical expressions (but don't do any calculations) for the following probabilities:
    (a) $P(X = 12)$,
    (b) $P(X \leq 2)$,
    (c) $P(32 < X < 35)$,
    (d) $P(X < 47)$.

11. Let X and Y be the number of successes and the proportion of successes respectively in 100 tries of a Bernoulli trial with $p = 1/4$. Find the expected value and variance of both X and Y.

12. Let X be a binomial random variable with $n = 10$ and $p = 0.7$. Determine the following probabilities:
    (a) $P(X < 4)$,

(b) $P(X \geq 3)$,

(c) $P(X = 6)$,

(d) $P(0 < X < 9)$.

13. Let Y be the random variable in Sample Problem 5 of the text. Compute $P(Y < .4)$ and $P(Y > .25)$.

14. Let X be a binomial random variable with $n = 600$ and $p = 2/5$. Calculate the standard deviation of X.

15. Explicitly write out the statement of the Law of Large Numbers with $k = 5$ for Y the proportion of times the outcome of $1,400$ rolls of a pair of fair dice is 7 or 11.

16. Let Y be the proportion of successes in 8 tries of a Bernoulli trial with $p = 0.35$. Find $P(Y > 1/2)$ and $P(Y < 2/3)$.

17. How many seeds should the laboratory in Sample Problem 6 of the text plant to be 96% certain that the proportion which germinate is between 0.78 and 0.82?

18. Let Y be the proportion of successes in 200 tries of a Bernoulli trial with $p = 1/3$. Compute the standard deviation of Y.

19. A wheel of chance used at a fair is numbered from 1 to 40 inclusive, and each number is divided into 3 equal parts. The center part is painted red and the outer two parts are painted white. When it stops on the red part, the payoff for anyone betting on that number is doubled. Suppose you want to verify that the wheel really stops on red 1/3 of the time. How many times would you have to spin the wheel to be 89% certain, i.e. $k = 3$, that the proportion of times the wheel stops on red would deviate from 1/3 by at most 0.01?

20. Use Table II to calculate the probability function for a random variable X which is $b(4, .35)$.

6.2 Normal Approximation

Let X be a binomial random variable. In the first section of this chapter it was shown that

$$P(X = k) = \binom{n}{k} p^k q^{n-k}$$

where p and q are the probabilities of success and failure in the underlying Bernoulli trial and n is the number of times it is repeated. Although this is an exact formula for $P(X = k)$, it is not very useful because the arithmetic quickly becomes horrendous. However, if we are willing to sacrifice exactness, there is a better way to compute a reasonably accurate estimate of $P(X = k)$ using the normal random variables discussed in Section 5.2.

To understand the relationship between binomial and normal random variables, it is necessary to compare the histogram of a binomial random variable with the probability density of a normal random variable. Figure 1 shows a histogram of a binomial random variable with $n = 10$ and $p = 0.5$.

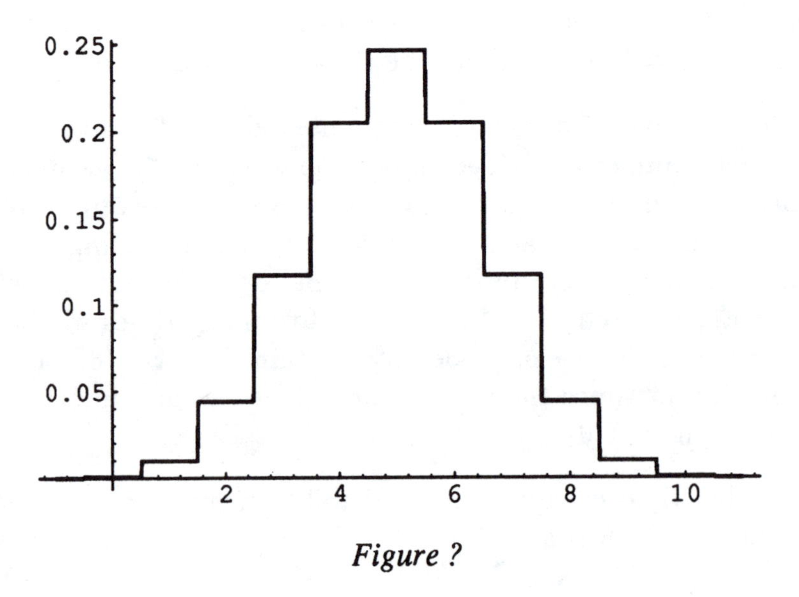

Figure ?

The vertical lines in earlier histograms like Figure 1 from the previous section have been deleted here. Without these vertical lines the similarities between a histogram and a probability density are more apparent.

For histograms unfortunately $P(X < a)$ does not always equal the area beneath the curve to the left of a. For example, when $a = 3$,

$$P(X < 3) = P(X = 0) + P(X = 1) + P(X = 2)$$

which is the area of the first three rectangles in Figure 1. The area beneath the curve to the left of 3 in Figure 1 is the area of the first three rectangles plus half the area of the fourth rectangle and does not equal $P(X < 3)$. However, when $a = 2.5 = 2 + 0.5$,

$$P(X < 2.5) = P(X = 0) + P(X = 1) + P(X = 2)$$

and the area beneath our curve to the left of 2.5 is the area of the first three rectangles which is precisely $P(X = 0) + P(X = 1) + P(X = 2)$. So the probability that X is less than 2.5 does equal the area beneath this curve to the left of 2.5. Similarly, any time $a = k+0.5$ or $a = k-0.5$, k an integer, $P(X < a)$ equals the area beneath the curve in Figure 3 to the left of a. For all other a's this is false. Therefore the curve in Figure 1 can be used as a probability density for the binomial random variable X provided we work only with numbers of the form $k + 0.5$ or

$k - 0.5)$, k and integer.

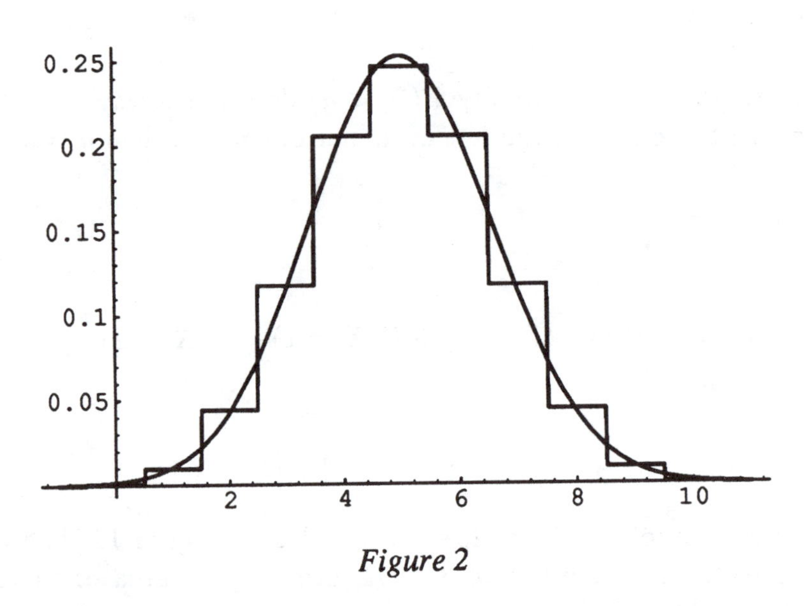

Figure 2

The key obsevation is that Figure 1 looks like a normal curve with corners. Figure 2 shows the probability density for a normal random variable

$$\sqrt{npq}Z + np$$

superimposed on Figure 1. Notice that the fit is quite good and that the corners protruding from the normal curve are counter balanced by missing pieces under the normal curve. When n is increased, the probability density for X constructed with rectangles looks more and more like a normal curve. In other words, X is approximately normally distributed. Since $\mu = np$ and $\sigma = \sqrt{npq}$ for X, this means the probabilities for $\sqrt{npq}Z + np$ are distributed in almost the same way as the probabilities for X with the error small when n is large. Thus the normal random variable $\sqrt{npq}Z + np$ can be used to calculate probabilities for the binomial random variable X. Although this method introduces a slight error in the calculations, it is a price worth paying because the probabilities for $\sqrt{npq}Z + np$ are easy to determine with the help of Table I.

Sample Problem 1. *A Bernoulli trial with probability of success 2/3 is repeated 450 times. Let X be the number of successes. Use the normal approximation to estimate the following probabilities:*

(a) $P(X \leq 306)$

(b) $P(X < 312)$

(c) $P(X > 320)$

(d) $P(X = 301)$.

SOLUTION: Before the normal approximation can be used, μ and σ must be determined. Because $p = 2/3, q = 1/3$ and $n = 450$,

$$\mu = 450 \cdot \frac{2}{3} = 300$$

and

$$\sigma = \sqrt{450 \cdot \frac{2}{3} \cdot \frac{1}{3}} = \sqrt{100} = 10.$$

We will now think of each of the probabilities to be estimated as an area beneath the histogram for X and find the corresponding area under the normal density for $10Z + 300$. This forces us to use numbers of the form $k + 0.5$ or $k - 0.5$ where k is an integer.

(a) The probability that $X \leq 306$ equals

$$P(X = 0) + P(X = 1) + \ldots + P(X = 306)$$

which is the area of all the rectangles to the left of $306 + 0.5 = 306.5$ Thus

$$\begin{aligned}
P(X \leq 306) &= P(X < 306.5) \\
&\approx P(10Z + 300 < 306.5) \\
&= A\left(\frac{306.5 - 300}{10}\right) \\
&= A\left(\frac{6.5}{10}\right) \\
&= A(0.65) = 0.7422.
\end{aligned}$$

(The symbol $\approx$ means approximately equal.)

The critical step in these problems is deciding which is the last (first) rectangle whose area must be included. Had we used $P(X < 305.5)$ the rectangle above 305.5 to 306.5 with area $P(X = 306)$ would not be included. This would have been wrong because 306 certainly satisfies the condition $X \leq 306$.

(b) The condition $X < 312$ excludes the value 312 and hence the rectangle whose area is $P(X = 311)$ should be the last rectangle included in the calculation. That is, $P(X < 312)$ equals the area beneath the probability density for X to the left of $312 - 0.5 = 311.5$. Now

$$\begin{aligned} P(X < 312) &= P(X < 311.5) \\ &\approx P(10Z + 300 < 311.5) \\ &= A\left(\frac{311.5 - 300}{10}\right) \\ &= A(1.15) = 0.8749. \end{aligned}$$

(c) The complement of $X > 320$ is $X \leq 320$ which implies

$$P(X > 320) = 1 - P(X \leq 320).$$

As in part (a)

$$\begin{aligned} P(X \leq 320) &= P(X < 320.5) \\ &\approx P(10Z + 300 < 320.5) \\ &= A\left(\frac{320.5 - 300}{10}\right) \\ &= A(2.05) = .9798 \end{aligned}$$

Combining the two equations P(X ¿ 320) = 1 - 0.9798 = 0.0202.

(d) The probability that X equals 301 is the area of the rectangle above the interval from 300.5 to 301.5. Thus

$$P(X = 301) = P(300.5 < X < 301.5)$$

$$\begin{aligned} &\approx P(300.5 < 10Z + 300 < 301.5) \\ &= A\left(\frac{301.5 - 300}{10}\right) - A\left(\frac{300.5 - 300}{10}\right) \\ &= A(0.15) - A(0.05) \\ &= 0.5596 - 0.5199 = 0.0397 \end{aligned}$$

which is infinitely easier than using the formula

$$P(X = 301) = \binom{450}{301} \cdot \left(\frac{2}{3}\right)^{301} \cdot \left(\frac{1}{3}\right)^{149}$$

In summary this solution shows in detail how Table I is used to calculate probabilities for binomial random variables. □

How accurate is this procedure? Applying the new method to a problem to which Table II can also be applied would at least give us an indication of the accuracy.

EXAMPLE 1. Suppose X is $b(9, 0.2)$, $\mu = 9 \cdot (0.2) = 1.8$ and

$$\sigma = \sqrt{9 \cdot (0.2) \cdot (0.8)} = \sqrt{1.44} = 1.2.$$

On the one hand, the normal approximation of $P(X \leq 2)$ is

$$\begin{aligned} P(X \leq 2) &= P(X < 2.5) \\ &\approx P(1.2Z + 1.8 < 2.5) \\ &= A\left(\frac{2.5 - 1.8}{1.2}\right) = A\left(\frac{0.7}{1.2}\right) \\ &= A(0.583\ldots) = 0.7190. \end{aligned}$$

On the other hand, from Table II $P(X \leq 2) = 0.7382$. Although the error is only 0.0192, it would have been even smaller had we used a p close to 1/2 and/or an n larger than 9.

Let X be a binomial random variable and as before let $Y = (1/n)X$ denote the proportion of successes. The fact that X is approximately normally distributed means that, ignoring small errors,

$$X = \sqrt{pqn}Z + np$$

If this equation is multiplied by $1/n$, we get

$$Y = \frac{1}{n}X = \frac{\sqrt{pqn}}{n}Z + \frac{np}{n}$$

or

$$Y = \sqrt{\frac{pq}{n}}Z + p$$

because

$$\frac{\sqrt{pqn}}{n} = \frac{\sqrt{pqn}}{\sqrt{n^2}} = \sqrt{\frac{pqn}{n^2}} = \sqrt{\frac{pq}{n}}.$$

Therefore, Y is also approximately normally distributed. Because the values of Y are between 0 and 1 inclusive, the adjustment by plus or minus 0.5 is not relevant. (*The only time we will ever add or subtract* 0.5 *is when we are doing a normal approximation of a binomial random variable.*)

Sample Problem 2. *The probability that a high school student who has been admitted to a certain college will actually enroll is* $2/5 = 0.4$. *If 600 students are offered admission, what is the probability that the proportion who enroll is between* 0.37 *and* 0.41*?*

SOLUTION: Think of each offer of admission as a Bernoulli trial and let success mean the student enrolls. Then $p = 2/5 = 0.4, q = 3/5 = 0.6$, and $n = 600$. Let Y be the proportion of successes. The problem is to calculate $P(0.37 < Y < 0.41)$, and the best way to proceed is to use the fact that Y is approximately normally distributed; i.e. assume

$$Y = \sqrt{\frac{pq}{n}}Z + p.$$

Once $\sqrt{pq/n}$ has been calculated, the technique from Section 5.2 can be applied. After pressing a few buttons on a calculator

$$\sqrt{\frac{pq}{n}} = \sqrt{\frac{0.4 \cdot 0.6}{600}} = 0.02.$$

For all practical purposes $Y = 0.02Z + 0.4$ and proceeding as usual with a normal random variable

$$P(0.37 < Y < 0.41) = P(0.37 < 0.02Z + 0.4 < 0.41)$$

$$\begin{aligned}
&= A\left(\frac{0.41-0.4}{0.02}\right) - A\left(\frac{0.37-0.4}{0.02}\right) \\
&= A\left(\frac{0.01}{0.02}\right) - A\left(\frac{-0.03}{0.02}\right) \\
&= A(0.5) - A(-1.5) \\
&= 0.6915 - (1 - 0.9332) \\
&= 0.6915 + 0.9332 - 1 = 0.6247.
\end{aligned}$$

to complete the solution. □

The previous example assumed that the probability of success p was known. A more natural problem is for p to be the unkown we wish to determine. We cannot hope to determine p precisely. But we can approximate p and compute the probability that our estimate for p meets a specified accuracy. To do this let Y be the proportion of successes in n tries of the Bernoulli trial. Since $E(Y) = p$, the numerical value obtained for Y by actually repeating the Bernoulli trial n times and computing the proportion of successes should be a good estimate for p. Furthermore, The Law of Large Numbers tells us that increasing n improves the accuracy of this estimate of p with a high degree of certainty. A careful statement as well as the method of solution for this kind of problem is presented in the next example.

Sample Problem 3. *A lumber company needs to know the probability p that a pine seedling planted in a cut area will survive for five years. They decide to conduct an experiment with some seedlings and use the proportion that survive for five years (successes) as an estimate for p. How many seedlings should they plant so that*

$$P(p - 0.01 < Y < p + 0.01) = 0.95?$$

In other words, they want to be 95% *certain that their experimental value is within* 0.01 *of the real p.*

SOLUTION: Planting a pine seedling is to be viewed as a Bernoulli trial and Y denotes the proportion of successes in n trials. The allowed deviation from the true p is specified as $d = 0.01$, and the goal is to determine n so that

$$P(p - 0.01 < Y < p + 0.01) \geq 0.95.$$

This problem could be done with the Law of Large Numbers as in the previous section, but using the normal approximation of Y will give a sharper less wasteful answer.

When Y is replaced by its normal approximation $\sqrt{pq/n}Z + p$,

$$P(p - 0.01 < Y < p + 0.01) = 0.95$$

becomes

$$P(p - 0.01 < \sqrt{pq/n}Z + p < p + 0.01) = 0.95.$$

Subtracting the mean p and dividing by the standard deviation $\sqrt{pq/n}$ produces

$$P\left(\frac{-0.01}{\sqrt{pq/n}} < Z < \frac{0.01}{\sqrt{pq/n}}\right) = 0.95.$$

This is a problem of the form

$$P(-t < Z < t) = 0.95.$$

Because $P(-t < Z < t) = 2A(t) - 1$, our problem can be written

$$2A\left(\frac{0.01}{\sqrt{pq/n}}\right) - 1 = 0.95.$$

When we add 1 to both sides and divide both sides by 2, we get

$$A\left(\frac{0.01}{\sqrt{pq/n}}\right) = \frac{1.95}{2} = 0.975.$$

From Table I $A(1.96) = 0.9750$ and it follows that

$$\frac{0.01}{\sqrt{pq/n}} = 1.96$$

or

$$0.01 = 1.96\sqrt{pq/n}.$$

By squaring both sides this becomes

$$(0.01)^2 = (1.96)^2 pq/n$$

and solving for n gives us

$$n = \frac{(1.96)^2 pq}{(0.01)^2}.$$

Because we do not know p or q this formula appears to be useless.

One way out of this difficulty is to use an n which we are sure is larger than

$$\frac{(1.96)^2 pq}{(0.01)^2},$$

since planting more seedlings than necessary will improve the result of the experiment. It is known that

$$pq = p(1-p) = p - p^2 \leq 1/4.$$

($p - p^2$ is a quadratic or parabola opening downward and its extreme point or maximum is $(1/2, 1/4)$.) Now replacing pq in the above expression by $1/4$ causes an increase and gives us the inequality

$$\frac{(1.96)^2 pq}{(0.01)^2} \leq \frac{(1.96)^2(1/4)}{(0.01)^2} = \frac{(1.96)^2}{4(0.01)^2} = 9604.$$

Therefore, by planting 9,604 seedlings the desired accuracy would certainly be achieved. Although we will take 9,604 as the correct answer, in practice the lumber company would more than likely plant 10,000 seedlings. However, had we carefully worked the same problem using only the Law of Large Numbers we would have gotten an answer of 50,000 not 10,000. Although the Law of Large Numbers guaranteed that repeating the experiment often enough would give the desired estimate of p, knowing that Y was approximately normally distributed greatly improved the efficiency of the experiment. □

This method can be summarized as follows: To find n large enough to guarantee that

$$P(p - d < Y < p + d) = r$$

first use Table I to find t so that

$$A(t) = \frac{r+1}{2}$$

and then take n to be at least

$$\frac{t^2}{4d^2}$$

In the above example r, the degree of confidence was 0.95, t was 1.96, and d, the allowed deviation from p was 0.01.

Suppose the lumber company in Example 3 carries out the experiment with 10,000 seedlings and 5835 of them survive for five years. Then the proportion of success or the value of Y would be

$$\frac{5835}{10{,}000} = 0.5835.$$

From this one statistic they would infer that p was about 0.58 and be 95% confident that their inference was correct. This brings us to the beginning of statistical inference and the end of these notes.

Exercises

1. Use the normal approximation to compute the following probabilities for the binomial random variable X from Example 1 of the text:

   (a) $P(X \leq 308)$,
   (b) $P(X < 315)$,
   (c) $P(X = 303)$,
   (d) $P(296 \leq X \leq 304)$,
   (e) $P(X < 290)$,
   (f) $P(X > 298)$.

2. The probability that a slightly defective soda machine will dispense a soda when properly operated is 3/5. During a day 150 people attempt to use this machine to purchase a soda. Use the normal curve approximation to estimate that (a) at most 100 succeed, (b) less than 75 succeed. (This is the same situation as Example 4 in 6.1 where we showed that $\mu = 90$ and $\sigma = 6$ for X equals the number of times the machine dispenses a soda.)

3. A fair die is rolled 720 times. Let X be the number of times the result is not a one. Use the normal curve approximation to estimate the following:

   (a) $P(X = 605)$,
   (b) $P(X \leq 621)$'
   (c) $P(X < 585)$,
   (d) $P(X \geq 625)$,
   (e) $P(580 \leq X \leq 610)$,
   (f) $P(X = 610)$.

4. Figure 2 in the text shows the approximation of a binomial random variable with distribution $b(10, 0.5)$ by the normal random variable $1.6Z + 5$. (The mean and standard deviation of X are

5 and $\sqrt{2.5} = 1.6$ respectively.) Use Table II to compute the area of the rectangle sitting above the interval from 5.5 to 6.5, i.e. $P(X = 6)$. Then use the normal approximation to estimate $P(X = 6)$. Compare the results.

5. Let Y be the random variable in Example 2. of the text. Use the normal approximation to estimate $P(Y < 0.35)$ and $P(Y > 0.42)$.

6. In a large grocery store, the probability that a customer will pay for an order with a check is 1/5. During the day 100 orders are rung up at one of the cash registers. Use the normal curve approximation to compute the probabilities that exactly 22 of these orders were paid for with a check and that at least 25 of these orders were paid for with a check. (For simplicity round numbers like 1.875 down to 1.87 before using Table I.)

7. A coin which has been altered so that the probability of heads is 0.55 is tossed 99 times. Use the normal curve approximation to estimate the probability that the proportion of heads is less than 0.52. (The numbers have been chosen so that the arithmetic works out nicely, and $(.05)^2 = 0.0025$ will be helpful.)

8. Let Y be the proportion of successes in 12 tries of a Bernoulli trial with $p = 0.4$. Thus the mean and standard deviation of Y are 0.4 and $\sqrt{0.02} = .14$. Use both Table II and the normal curve approximation to calculate $P(Y < 0.45)$ and compare your answers.

9. An Eastern Shore waterman puts out 48 crab pots. He knows when he returns to check them the probability of finding crabs in a pot is 1/4. Use the normal curve approximation to estimate the probability that he finds crabs in at most 8 pots. Same question for at least 14 but no more than 18.

10. How many seedlings should the lumber company in Example 3 of the text plant so that

    (a) $P(p - 0.05 < Y < p + 0.05) = 0.95$

    (b) $P(p - 0.01 < Y < p + 0.01) = 0.98$

(c) $P(p - 0.02 < Y < p + 0.02) = 0.90$?

11. Before deciding on the cost of a comprehensive two year service contract for a new television, the manufacturer would like to know the probability p that the picture tube will not burn out in less than two years. They decide to conduct an experiment with some sets and use the proportion of successes Y as an estimate for p. How many sets should they test so that $P(p-0.1 < Y < p+0.1) = 0.80$?

12. A pair of fair dice are rolled 288 times. Let X be the number of times the sum of the outcomes is 4, 7 or 10. Calculate the following probabilities:

    (a) $P(X \leq 100)$,

    (b) $P(90 < X < 102)$,

    (c) $P(X < 95)$.

13. Suppose the Social Security Administration wants to know the probability p that a random member contributed less than $1,000 to the system in 1981. Using the computer they can select a member at random, call up his/her record, and look at the amount contributed to Social Security in 1981. How many times would they have to repeat this to be 85% confident that the proportion who contributed less than$1,000 deviated from p by at most 0.03?

14. In a gambling case the district attorney suspects that he has a pair of loaded dice which can be used as evidence. He wants to tell the jury the actual probability of rolling a 7 or 11 with this pair of dice. Moreover, he wants to be 99% confident that he is correct to within 0.01. How many times would he have to roll this pair of dice to get the information he wants?

15. Let X be a binomial random variable with $n = 12$ and $p = 1/4$. Use the normal curve approximation to estimate $P(X \leq 3), P(X \leq 4)$ and $P(X \leq 6)$. Then use Table II to determine the error in each estimate.

6.3 Review Problems

1. The random variable X is the number of successes in a Bernoulli trial where there are 98 trials and the probability of success on a single trial is 1/7. Find the expected value, the variance and the standard deviation of X.

2. A Bernoulli trial with $p = 7/20$ is repeated 9 times. Let X be the number of successes. Determine

   (a) $P(X \geq 4)$,

   (b) $P(X > 2)$,

   (c) $P(1 < X < 8)$,

   (d) $P(X = 5)$.

3. The probability that a certain kind of tomato plant will survive transplanting at a certain age is 3/4. If a farmer transplants 48 of them, using the normal curve approximation estimate the probability that at most 31 will survive.

4. Let X be a binomial random variable with $n = 11$ and $p = .55$. Determine the following probabilities:

   (a) $P(X = 6)$,

   (b) $P(X \leq 4)$,

   (c) $P(X \geq 8)$,

   (d) $P(3 \leq X \leq 9)$.

5. You have an unfair coin and you want to determine the probability p of heads. You do this by flipping it several times and finding the proportion of heads obtained. How many times should you flip the coin in order to be 95.44% sure that your value of p is correct to within 0.01?

6. A fair eight sided (octahedron) die is rolled 10 times. What is the probability that less than two times the outcome will be a multiple of 3? What is the probability that more than 2 times the outcome will be a multiple of 3?

7. Suppose a pair of dice are rolled 1,100 times and X is the number of times the sum of the faces is 4. Compute $E(X)$, $V(X)$ and $\sigma(X)$.

8. The probability that any particular can of tennis balls produced by a certain company is defective (a can is defective if it contains at least one defective tennis ball) is 1/15. Yesterday they produced 3150 cans of tennis balls. Let X be the number of defective cans produced yesterday. Compute the expected value and standard deviation of X. (Hint: the latter is an integer.) Use the normal curve approximation to estimate the following probabilities:

   (a) They produce at most 227 defective cans.

   (b) They produce at least 214 defective cans.

9. An urn contains a mixture of red and green balls. You may randomly select one ball, note its color, and return it to the urn; but you may not look into the urn. How many times would you have to randomly select a ball to determine the proportion of red balls in the urn with an accuracy of 0.05 and be 92% confident of your answer?

10. Let X be a binomial random variable with mean 24 and standard deviation 4. Estimate $P(25 < X < 30)$.

11. A company manufacturing flash bulbs wants to determine the probability p that one of their bulbs will not fire properly when used on a camera with good batteries. How many bulbs must they test to get an estimate for p which they are 98% confident deviates from p by at most .001?

12. A Bernoulli trial with $p = 2/3$ is repeated 162 times. Compute the variance and the standard deviation for the proportion of successes.

13. A coin which has been altered so that the probability of heads is 3/4 is tossed 4 times. Let X be the number of heads and compute its probability function.

14. The probability that a new car will not start when it rolls off a certain assembly line is 2/5. If on a given day 600 cars roll off this assembly line, what is the probability that

    (a) less than 251 of them will not start?

    (b) the number of them that will not start is more than 236 and less than 244?

15. In an electronic game the computer fires missiles at your space ship. The missiles randomly come from any point along the bottom of the screen. When they come from the middle 3/5 they are easier to evade. What is the probability that less than 5 out of 12 missiles will come from the middle 3/5 of the bottom of the screen? What is the probability that out of 12 missiles at least 6 but no more than 9 will come from the middle 3/5 of the bottom of the screen?

16. Suppose that a machine making yo-yo's has a 10% chance of producing a defective yo-yo each time it makes one. Let X be the number of defective yo-yo's in a shipment of 400. Estimate $P(X \geq 33)$.

17. Answer the following questions for a loaded die given the probability that the outcome is 6 equals 1/5:

    (a) What is the probability that the outcome is 6 on at least 3 of 9 rolls of this die?

    (b) If this die is rolled 625 times, what is the exact probability that the outcome is 6 exactly 127, 128 or 129 times? (Don't compute.)

    (c) What is a good numerical estimate for your answer to b)?

18. A pair of dice are rolled 180 times. Let X be the number of times doubles are rolled. Use the normal curve approximation to estimate the probability of getting exactly 35 doubles.

19. A spinner in a child's board game is numbered from 1 to 10 and each number is supposed to be equally likely. When the spinner stops at 8 you must move backwards 8 spaces.

(a) During the course of one game the spinner is used 100 times. Calculate the probability that the proportion of times it stops at 8 is greater than 0.12.

(b) After playing the game you suspect the spinner is biased in favor of 8. How many times would you have to spin it to be 80% confident that the proportion of times it stops at 8 deviates from the probability of 8 by at most 0.01?

20. The probability that a certain coin operated copier will fail to feed in a blank sheet of paper is 1/5 each time that it is used. A graduate student uses this machine to make a copy of his 100 page thesis. If he does not check for missing pages until he has put each of the 100 pages in the machine once, how many missing pages can he expect to find? What is the standard deviation for the number of missing pages? Use the normal curve approximation to estimate the following probabilities:

    (a) there are more than 24 missing pages,

    (b) there are exactly 20 missing pages.

21. A bottling machine at a dairy occasionally fails to properly cap a bottle. If the dairy knew the probability p that this machine will not properly cap a bottle, they could use it to decide their course of action on this problem. How many bottles coming from this machine would they have to check to be 75% confident that the proportion of improperly capped bottles deviated from p by at most 0.02?

22. A candidate for president wants to find out what his chances are in the New Hampshire primary. He hires a pollster to determine the probability p that a randomly selected voter from New Hampshire will vote for him. (If, for example, $p = .59$, then he would expect to win 59% of the popular vote.) The candidate requires the pollster to produce an estimate for p that he is 80% confident is within 3% of p. The pollster charges $5.00 per person in the sample. What is the candidate's minimal cost for this work?

23. Let X be the number of times out of 180 throws of a fair die that the outcome is a 4. Let A be the event $25 \leq X \leq 35$ and B the event that $X > 30$. Use the normal curve approximation to estimate $P(A|B)$.

24. Problem 19 in Section 6.1 can also be solved using the techniques from Section 6.2. This problem was concerned with a wheel of chance on which each number was divided into 3 equal parts, one red and two white. Let p equal the probability of the outcome being a red number. Use the formula at the end of Section 6.2 to determine how many times you would have to spin this wheel to be 89% confident that the proportion of successes deviates from p by at most .01. (If the proportion of successes was within .01 of $1/3$, then we would be 89% confident the wheel was fair.) Compare your answer with the answer to the original problem. Which method is more efficient?

25. Let X be the number of eggs eaten per week by an adult living in the U.S.. Describe how you would obtain an accurate estimate of the probability function of X for a national egg producer's association.

Appendix A

Answers to Odd Exercises

A.1 Chapter 1

Section 1

1. (a) $A \subset B$, (b) neither (c) $C \subset B$

3. (a) $\{3, 4, 5, 8\}$, (b) $\{3, 5\}$ (c) $\{1, 2, 3\}$, (d) $\{7\}$.

5. (a) $\{4, 5\}$ (b) $\{1, 3, 4, 5\}$, (c) $\{1, 5\}$.

7. (a) $\{1, 3, 4, 5, 6, 7, 9\}$, (b) $\{3, 4, 9\}$, (c) $\{2, 6, 7, 8, 10\}$, (d) $\{3, 4, 9\}$.

9. (a) $\{3, 5, 11\}$, (b) $\{5, 9, 13\}$, (c) $\{1, 3, 7, 9, 13\}$.

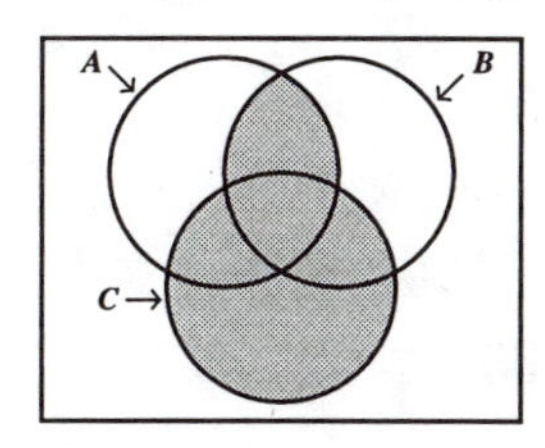

11.

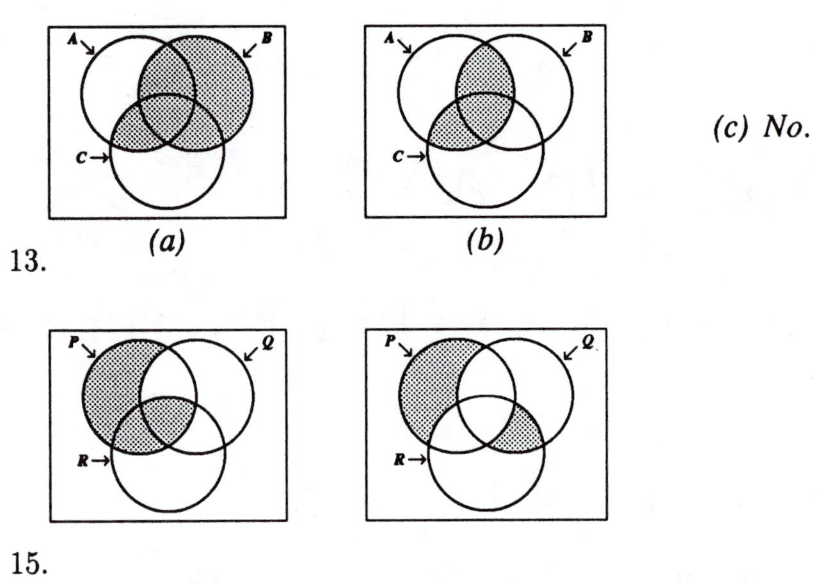

(a) (b)

13. (c) No.

15.

17 No.

Section 2

1. $\{L, T, V, X\}$, $\{A, F, H, I, K, N, Y, Z\}$ and $\{E, M, W\}$.

3. $\{5, 7, 11, 13\}$, $\{4, 6, 9, 10, 14, 15\}$, $\{8, 12\}$, and $\{16\}$.

5. (a) 41, (b) 13, (c) 19, (d) 49.

7. (a) 148, (b) 53, (c) 14.

9. (a) 2, (b) 18, (c) 37.

11. (a) 50, (b) 105, (c) 45.

13. (a) 40, (b) 85, (c) 5.

15. 16 and 48.

17. (a) 54, (b) 22, (c) 35.

19. 7, 63.

21. 5.

23. (a) 19, (b) 10, (c) 40.

Section 3

1. $S = \left\{ \begin{array}{l} (H,1),(H,2),(H,3),(H,4),(H,5),(H,6), \\ (T,1),(T,2),(T,3),(T,4),(T,5),(T,6) \end{array} \right\}.$

   (a) $\{(H,1),(H,3),(H,5)\}$.

   (b) $\{(T,1),(T,2),(T,3),(T,4),(T,5),(T,6),(H,2),(H,4),(H,6)\}$.

   (c) $\{(T,1),(T,2),(T,3),(T,4),(T,5)\}$.

3. $S = \{AB, AC, AD, BA, BC, BD, CA, CB, CD, DA, DB, DC\}$ where the first letter denotes the friend getting the ticket on the 50 yard line. $E = \{AB, AC, AD, DA, DB, DC\}$.

5. $S = \{ab, ac, ad, ba, bc, bd, ca, cb, cd, da, db, dc\}$

7. $S = \left\{ \begin{array}{l} 1248, 1284, 1428, 1482, 1824, 1842, \\ 2148, 2184, 2418, 2481, 2814, 2841, \\ 4128, 4182, 4218, 4281, 4812, 4821, \\ 8124, 8142, 8214, 8241, 8412, 8421 \end{array} \right\}$

Section 4

1. 0.006

3. 0.1

5. 2/3, 1/3, and 2/5.

7. (a) 17/20, (b) 3/20, (c) 13/20, (d) 1/20.

9. (a) and (c) are impossible

11. 0.25, 0.45, and 0.9.

13. (a) 1/10, (b) 1/4, (c) 9/10, (d) 1/2.

15. (a) 0.31, (b) 0.23, (c) 0.12.

17. (a) 0.20, (b) 0.29, (c) 0.43.

19. (a) 0.51, (b) 0.88, (c) 0.45, (d) 0.17.

21. (a) 0.4, (b) 0.5.

Section 5

1. $S = \left\{ \begin{array}{c} (1,1),(1,2),(1,3),(2,1),(2,2),(2,3), \\ (3,1),(3,2),(3,3),(4,1),(4,2),(4,3) \end{array} \right\}$. (a) 1/2, (b) 1/6, (c) 1/4, (d) 3/4.

3. (a) 1, (b) 0.05, (c) 0.1, (d) 0.3.

5. (a) 1/4, (b) 5/8, (c) 7/16.

7. 3/10.

9. 1/4 and 3/8.

11. 0.29.

13. (a) .35, (b) .7.

15. 9.

17. 5:1 and 7:2.

19. 4/7.

A.2 Chapter 2

Section 1

1. 28.

3. 48.

5. (a) 16/81, (b) 65/81, (c) 32/81, (d) 8/81, (e) 80/81.

7. $26^4 \cdot 5 \cdot 10^2 = 228,488,000$

9 1,800 and 1,290.

11. 110.

13. 525, 1,575, and 2,730.

15. 6^{10}.

17. $24^2 + 24^3 = 14,400$ and $24 \cdot 23 + 24 \cdot 23 \cdot 22 = 12,696$.

19. $26^3 + 26^4 + 26^5$.

Section 2

1. (a) 120, (b) 840, (c) 40, (d) 600.

3. $\mathbf{p}(6,3) = 120$, $\mathbf{p}(5,2) = 20$, $2 \cdot \mathbf{p}(5,2) = 40$.

5. $12! = 479,001,600$ and $6! \cdot 2^6 = 46,080$.

7. (a) 12!, (b) $4! \cdot 8!$, (c) $\mathbf{p}(12,9) = 79,833,600$.

9. 8!, $12 \cdot 6! = 8,640$.

11. $\mathbf{p}(15,9) = 1,816,214,400$ and $\mathbf{p}(5,3)\mathbf{p}(12,6) = 39,916,800$.

13. $\mathbf{p}(11,2) = 110$.

15. 9/13, 235/676, and 1/100.

17. 1/6.

19. (a) 720, (b) 240, (c) 480, (d) 144.

Section 3

1. (a) 210, (b) 190, (c) 56, (d) 27, (e) 5050, (f) 85.

3. 1330.

5. 455, 364.

7. $\binom{11}{5} = 462$ and $p(6,5) = 720$.

9. $\frac{\binom{8}{2}}{\binom{50}{2}} = \frac{28}{1225} = 0.0228571.$

11. (a) $\frac{\binom{5}{1}^4}{\binom{20}{4}} = 0.128999$ (b) $\frac{\binom{4}{2}\binom{5}{2}^2}{\binom{20}{4}} = 0.123839$
(c) $\frac{\binom{5}{1}\binom{4}{3}\binom{15}{1}+\binom{5}{1}\binom{4}{4}}{\binom{20}{4}} = 0.062951.$

13. $\frac{\binom{20}{2}^3}{\binom{60}{6}} = 0.137005,$ $\frac{\binom{3}{2}\binom{20}{3}^2}{\binom{60}{6}} = 0.77877,$ and $\frac{\binom{3}{1}\binom{20}{5}\binom{40}{1}+\binom{3}{1}\binom{20}{6}}{\binom{60}{6}} = 0.039485.$

15. $\frac{\binom{12}{2}\binom{24}{4}}{\binom{36}{6}} = 0.360057$ and $1 - \frac{\binom{24}{6}}{\binom{36}{6}} = 0.930898.$

17. $\binom{9}{3}\binom{6}{2}4! = 30,240,$ $\binom{8}{3}\binom{5}{2}\binom{3}{2} = 1,680,$ $\binom{8}{3}5! = 6,720,$ and $\binom{6}{2}\binom{4}{2}2! = 180.$

19. $\binom{18}{6}\binom{12}{5}\binom{7}{4}3! = 3,087,564,480.$

21. $\binom{8}{3}\binom{5}{2}3! = 3,360.$

23. $\frac{\binom{8}{1}\binom{12}{5}+\binom{8}{2}\binom{12}{4}+\binom{8}{3}\binom{12}{3}}{\binom{20}{6}} = 0.84.$

25. (a) $\frac{\binom{4}{1}\binom{26}{6}}{\binom{104}{6}} = 0.000607,$ (b) $\frac{\binom{26}{1}\binom{4}{4}\binom{100}{2}}{\binom{104}{6}} = 0.000085,$
(c) $\frac{\binom{26}{2}\binom{4}{3}^2}{\binom{104}{6}} = 0.000003.$

27. 75,287,520.

29. 1/13, 3/13, and 8/13.

31. (a) 480, (b) 368.

33. (a) $\frac{4}{\binom{52}{5}} = 0.00000154,$
(b) $\frac{36}{\binom{52}{5}} = 0.0000138,$

(c) $\frac{\binom{13}{1}\binom{4}{4}\binom{48}{1}}{\binom{52}{5}} = \frac{13 \cdot 48}{\binom{52}{5}} = 0.000240,$

(d) $\frac{\binom{13}{1}\binom{4}{3}\binom{12}{1}\binom{4}{2}}{\binom{52}{5}} = \frac{13 \cdot 4 \cdot 12 \cdot 6}{\binom{52}{5}} = 0.00144,$

(e) $\frac{\binom{4}{1}\binom{13}{5}-40}{\binom{52}{5}} = 0.00197,$

(f) $\frac{10\binom{4}{1}^5-\binom{4}{1}10}{\binom{52}{5}} = 0.00392,$

(g) $\frac{\binom{13}{1}\binom{4}{3}\binom{12}{2}\binom{4}{1}^2}{\binom{52}{5}} = 0.0211,$

(h) $\frac{\binom{13}{2}\binom{4}{2}^2\binom{11}{1}\binom{4}{1}}{\binom{52}{5}} = 0.0475,$

(i) $\frac{\binom{13}{1}\binom{4}{2}\binom{12}{3}\binom{4}{1}^3}{\binom{52}{5}} = 0.423.$

35. $\frac{\binom{32}{13}}{\binom{52}{13}}$ and $\frac{\binom{4}{1}\binom{13}{7}\binom{13}{2}^3}{\binom{52}{13}}$.

Section 4

1. $(x + 3)^4 = x^4 + 12x^3 + 54x^2 + 108x + 81$ and $(s-t)^7 = s^7 - 7s^6t + 21s^5t^2 - 35s^4t^3 + 35s^3t^4 - 21s^2t^5 + 7st^6 - t^7$.

3. $\binom{9}{6} = 84$, $\binom{6}{3} \cdot 3^3(-2)^3 = -4,320$.

5. $2^5 = 32$.

7. $2^6 - 1 = 63$.

9. $2^6 - 1 = 63$.

11. 57.

A.3 Chapter 3

Section 1

1. 1/3, 3/16, 2/5.

3. 1/2

5. 0.3

7. 28/55.

9. 1/8, 1/3.

11. (a) 6/11 = 0.5454 (b) 2/3 = 0.6666

Section 2

1. 15/32, 5/24, and 31/96.

3. 0.4

5. 0.25.

7. 0.4

9. 5/12 and 5/7.

11. 2/3

13. 7/20 and 3/14.

15. 30/32

17. (a) 0.24, (b) 0.25.

19. 3/4 = 0.75.

21. 2.48% and 19.76%.

Section 3

1. A and B are independent as are B and C. A and C are dependent.

3. (a) 0.27, (b) 0.93, (c) 0.63.

5. (a) 0.44, (b) 0.48, (c) 0.08.

7. (a) 0.092, (b) 0.098.

9. (a) 1/60, (b) 2/5, (c) 1/5.

11. (a) 0.24, (b) They are independent, (c) $0.4/0.76 = 0.526$.

13. 1/3

15. (a) 0.2, (b) 0.45, (c) 0.35.

17. (a) 0.992, (b) 0.012, (c) 0.124.

Section 4

1. $\binom{45}{10}(2/5)^{10}(3/5)^{35}$.

3. 21/128 and 29/128.

5. 0.00045.

7. (a) $\binom{21}{12}(1/3)^{12}(2/3)^{9}$, (b) $1-\left[\binom{21}{0}(1/3)^{0}(2/3)^{21} + \binom{21}{1}(1/3)^{1}(2/3)^{20}\right]$.

A.4 Chapter 4

Section 1

1. The values of X are 0, 1, 2, 3, 4. The partition is: $\{TTTT\}$ $\{TTTH, TTHT, THTT, HTTT\}$ $\{TTHH, THTH, HTTH, THHT, HTHT, HHTT\}$ $\{THHH, HTHH, HHTH, HHHT\}$ $\{HHHH\}$
The probability function is: $P(X = 0) = 1/16$, $P(X = 1) = 1/4$, $P(X = 2) = 3/8$, $P(X = 3) = 1/4$, $P(X = 4) = 1/16$.

3. $P(X = 0) = P(X = 3) = P(X = 4) = 1/6$, $P(X = 1) = P(X = 2) = 1/4$.

5. $a = 0.45$.

7. 80/81 and 1/9.

9. $P(X = 0) = 1/6, P(X = 1) = 5/9, P(X = 2) = 5/18$.

11. (a) 0.45, (b) 0.8, (c) 0.6.

13. $\{3, 4, 5, 6, 7, 8, 9, 10, 11, 12, 13, 14, 15, 16, 17, 18\}$,
    10/216, 15/216, 21/216.

15. $P(X = 1) = 3/4$, $P(X = 2) = 3/14$, $P(X = 3) = 1/28$

Section 2

1. 2

3. 6

5. $-1/5$

7. 4

9. 13

11. \$2.00.

13. The house by \$0.15, \$0.85.

15. \$10.00

17. 9 cents.

19. -25 cents.

Section 3

1. $\mu = 2$ and $\sigma^2 = 3.2$.

3. $\mu = 3$ and $\sigma^2 = 2$.

5. (a) 4, (b) 8, (c) 36, (d) 14, (e) 10, (f) 4/9.

7. 3

9. $E(Y) = .7$, $V(Y) = 2.61$, $E(X) = -8.9$, $V(X) = 23.49$.

11. $P(10 < X < 40) \geq 24/25 = 0.96$,
    $P(10 < X < 40) \geq 21/25 = 0.84$.

13. 0.91, 0.64, 0.19.

15. 7.

A.5 Chapter 5

Section 1

1. (a) 3/4, (b) 5/8, (c) 0, (d) 1/8.
3. $c = 2$, 1/4, and 3/4.
5. 0.3 and 0.8.
7. (a) 4, (b) 2.6, (c) 3.5.
9. 1/5 and 2/5.
11. 7/10.
13. (a) 1/4, (b) 3/4, (c) 1/16, (d) 15/16, (e) 7/16.
15. 1/20.

Section 2

1. (a) 0.9744, (b) 0.2843, (c) 0.5359, (d) 0.2054.
3. (a) 0.0418, (b) 0.9989, (c) 0.4116, (d) 0.9328.
5. (a) 1.66, (b) 0.13, (c) 0.27, (d) −1.26.
7. (a) 0.9554, (b) 0.3821, (c) 0.8664, (d) 0.1151,
9. (a) 46.44, (b) 40.08, (c) 49.53, (d) 41.04,
11. 0.1151 and 0.5762.
13. (a) 0.3373, (b) 0.6915.
15. (a) 34.8, (b) 21.85, (c) 24.35.
17. (a) 92.26, (b) 87.58.
19. 0.1056 and 4.656 ounces.
21. 8
23. (a) 0.2266, (b) 0.3944, (c) 13.12.
25. 89.44%.

A.6 Chapter 6

Section 1

1. $P(X=0)=0.0256$, $P(X=1)=0.1536$, $P(X=2)=0.3456$, $P(X=3)=0.3456$, $P(X=4)=0.1296$.

3. $P(X\le 0)=0.0256$, $P(X\le 1)=0.1792$, $P(X\le 2)=0.5248$, $P(X\le 3)=0.8704$, $P(X\le 4)=1$.

5. (a) 0.5941, (b) 0.9502, (c) 0.8936, (d) 0.2322.

7. (a) 0.0705, (b) 0.7443, (c) 0.3026, (d) 0.5791.

9. (a) $\binom{35}{17}(0.7)^{17}(0.3)^{18}$, (b) $\binom{35}{34}(0.7)^{34}(0.3)^{1}+\binom{35}{35}(0.7)^{35}(0.3)^{0}$, (c) $1-\left[\binom{35}{0}(0.7)^{0}(0.3)^{35}+\binom{35}{1}(0.7)^{1}(0.3)^{34}+\binom{35}{2}(0.7)^{2}(0.3)^{33}\right]$.

11. $E(X)=25$, $V(X)=75/4$, $E(Y)=1/4$, $V(Y)=3/1,600$.

13. 0.8424 and 0.3512.

15. $P(3/18<Y<5/18)\ge .96$.

17. 10,000.

19. 20,000.

Section 2

1. (a) 0.8023, (b) 0.9265, (c) 0.0381, (d) 0.3472, (e) 0.1469, (f) 0.5596.

3. (a) 0.0352, (b) 0.9842, (c) 0.0606, (d) 0.0071, (e) 0.8329, (f) 0.0242.

5. 0.0062 and 0.1587.

7. 0.2743.

9. 0.1210 and 0.2935.

11. At least 41.

13. At least 576.

15. 0.0195, 0.0011 and 0.0044.

Appendix B

Review Problem Answers

B.1 Chapter 1

1. 11

2. 0.95 and 0.85

3. (a) 42, (b) 33, (c) 18.

4. 86, 74.

5. (b) 6, (c) 35, (d)11.

6. $\{2,3,5,7\}, \{4,9\}, \{6,8,10\}$.

8. (a) $\{B_1\&B_2, B_1\&W_1, B_1\&W_2, B_2\&W_1, B_2\&W_2, W_1\&W_2\}$, (b) 2/3, (c) 1/3.

9. (a) 13, (b) 12, (c) 19.

10. (a) $\{u,v\}$, (b) $\{t,u,w\}$.

11. (a) 1, (b) 21.

12. 7/12, (b) 5/6, (c) 5/12.

13.
$$\{TTTT\} \cup \{HTTT, THTT, TTHT, TTTH\} \cup$$

$$\{HHTT, HTHT, THHT, HTTH, THTH, TTHH\} \cup$$
$$\{HHHT, HHTH, HTHH, THHH\} \cup \{HHHH\} = S$$

1/16, 4/16, 6/16, 4/16, 1/16.

14. Venn diagram

15. $\{2, 3, 7\}, \emptyset$.

16. (a) 0.1, (b) 0.24, (c) 0.35.

17. (a) $S = \left\{ \begin{array}{l} PN, PD, PQ, PH, NP, ND, NQ, NH, DP, DN, \\ DQ, DH, QP, QN, QD, QH, HP, HN, HD, HQ \end{array} \right\}$, (b) 1/2, (c) 3/20.

18. (a) 0.5, (b) 0.15, (c) 0.4.

19. No.

20. (a) 2/50, (b) 12/50, (c) 11/50.

21. (a) 64, (b) 1/64, (c) 1/4, (d) 3/32.

22. $S = \left\{ \begin{array}{l} (B1, G1), (B1, B2), (B1, G3), (B2, G1), (B2, B2), \\ (B2, G3), (G3, G1), (G3, B2), (G3, G3) \end{array} \right\}$, (a) 4/9, (b) 1/3, (c) 5/9, (d) 2/3, e) 1/3, f) 7/9.

23. (a) $\{1, 3, 8\}$, (b) $\{1, 2, 3, 8, 9\}$.

24. $S = \left\{ \begin{array}{l} LG\&LY, LG\&LB, LG\&RY, LG\&RB, RG\&LY, \\ RG\&LB, RG\&RY, RG\&RB, LY\&LB, LY\&RB, \\ RY\&LB, RY\&RB, LG\&RG, LY\&RY, LB\&RB \end{array} \right\}$ and 0.2 and 0.4.

25. $E' \cup (F \cap G)$ does equal $(E' \cup F) \cup (E' \cup G)$.

26. 8

27. (a) 1/10, (b) 17/50, (c) 29/50.

28. (a) 1/9, (b) 5/9, (c) 1/3, (d) 1, e) 0.

29 (a) 8, (b) 49, (c) 36, (d) 1/22 and 2/11.

B.2 Chapter 2

1. 14,400 and 64,800.

2. (a) 1/6, (b) 2/3, (c) 1/20.

3. $\mathbf{p}(9,6)$ and $2\mathbf{p}(5,3)\mathbf{p}(4,3)$.

4. (a) $\frac{\binom{60}{4}\binom{40}{8}}{\binom{100}{12}}$, (b) $\frac{\binom{40}{12}+\binom{60}{1}\binom{40}{11}}{\binom{100}{12}}$, (c) $\frac{\binom{60}{5}\binom{40}{7}+\binom{60}{6}\binom{40}{6}+\binom{60}{7}\binom{40}{5}}{\binom{100}{12}}$, (d) $1-\frac{\binom{40}{12}}{\binom{100}{12}}$.

5. (a) 258, (b) 156.

6. 30.

7. $\mathbf{p}(7,3)\binom{11}{5}$ and $\mathbf{p}(7,3)\binom{11}{5}+\mathbf{p}(7,2)\binom{11}{5}$.

8. $\frac{24}{\binom{20}{3}\binom{17}{6}\binom{11}{4}}$

9. (a) 15^4+15^5, (b) $\mathbf{p}(15,4)+\mathbf{p}(15,5)$, (c) 15^3+15^4, (d) $15\mathbf{p}(14,2)+15\mathbf{p}(14,3)$.

10. (a) $\frac{\binom{4}{2}\binom{14}{3}^2}{\binom{56}{6}}$, (b) $\frac{\binom{14}{2}\binom{4}{3}^2+\binom{14}{1}\binom{4}{4}\binom{13}{1}\binom{4}{2}}{\binom{56}{6}}$, (c) $\frac{\binom{14}{2}\binom{4}{2}^2\binom{12}{2}\binom{4}{1}^2}{\binom{56}{6}}$.

11. (a) $\frac{2\cdot 5!\cdot 5!}{10!}=0.0079$, (b) $\frac{6\cdot 5!\cdot 5!}{10!}=0.024$.

12. $\frac{\binom{15}{2}^3}{\binom{45}{6}}$ and $\frac{\binom{3}{2}\binom{15}{3}^2}{\binom{45}{6}}$, $1-\frac{\binom{15}{2}^3}{\binom{45}{6}}$.

13. $\mathbf{p}(7,5)=2,520$ and $\mathbf{p}(7,5)-\mathbf{p}(5,5)=2,400$.

14. $\frac{\binom{12}{3}\binom{18}{2}}{\binom{30}{5}}$, $\frac{\binom{18}{5}+\binom{12}{1}\binom{18}{4}+\binom{12}{2}\binom{18}{3}}{\binom{30}{5}}$.

15. (a) -792, (b) 84, (c) 66, (d) 1.4641.

16. (a) $\binom{15}{8}\binom{11}{7}\binom{9}{5}$, (b) 20!, (c) $\mathbf{p}(8,2)\mathbf{p}(18,18)$, (d) $\mathbf{p}(7,3)$.

17. (a) $\binom{28}{7}$, (b) $\frac{\binom{26}{5}+2\binom{26}{6}}{\binom{28}{7}}$ or $\frac{2\binom{27}{6}-\binom{26}{5}}{\binom{28}{7}}$, (c) $\binom{13}{2}+13=91$.

18. (a) $\frac{21\cdot 26^2\cdot 5}{26^4}$, (b) $\frac{\mathbf{p}(26,4)}{26^4}$, (c) $\frac{6\cdot 25\cdot 21^2}{26^4}$, (d) $\frac{6\cdot\mathbf{p}(21,2)\cdot 25+4\cdot 5\cdot\mathbf{p}(21,3)+\mathbf{p}(21,4)}{26^4}$, (e) $\frac{5\cdot 26^3+5\cdot 26^3-5^2\cdot 26^2}{26^4}$.

19. $(2\cdot 3)^8 = 6^8$.

20. $6,930$.

22. (a) $\binom{40}{7}$, (b) $\frac{\binom{18}{2}\binom{22}{5}}{\binom{40}{7}}$, (c) $\frac{\binom{2}{1}\binom{38}{6}+\binom{38}{5}}{\binom{40}{7}} = 1 - \frac{\binom{38}{7}}{\binom{40}{7}}$, (d) $\frac{\binom{20}{7}+\binom{18}{7}}{\binom{40}{7}}$, (e) $\frac{\binom{18}{6}\binom{22}{1}+\binom{18}{7}}{\binom{40}{7}}$.

23. (a) $\mathbf{p}(7,4)$, (b) $5\mathbf{p}(6,2)$, (c) $4\mathbf{p}(6,2)$.

24. $\binom{12}{4}\binom{8}{4}\binom{4}{4} = 34,650$.

25. $1 - \frac{\binom{3}{2}}{\binom{8}{2}} = 1 - \frac{3}{28} = 0.8928571$.

26. (a) $\binom{4}{1}\binom{10}{4}$, (b) $\binom{4}{1}\binom{10}{2}\binom{3}{2}\binom{10}{1}^2$, (c) $\frac{\binom{10}{1}^4}{\binom{40}{4}}$.

27. (a) 6^4, (b) $\mathbf{p}(6,4)$, (c) $\binom{4}{2}2!$, (d) $5^4 + \binom{4}{1}5^3$.

28. $\binom{11}{4}\binom{7}{3}\binom{4}{2}\binom{2}{2}$.

29. (a) $\frac{\binom{36}{6}\binom{12}{2}}{\binom{48}{8}}$, (b) $\frac{\binom{36}{3}\binom{12}{5}+\binom{36}{4}\binom{12}{4}+\binom{36}{5}\binom{12}{3}}{\binom{48}{8}}$.

30. $\mathbf{p}(5,2)\mathbf{p}(21,3)$ and $\binom{5}{3}\mathbf{p}(5,3)\mathbf{p}(21,2)$.

31. (a) $\frac{\binom{10}{6}+\binom{15}{6}+\binom{20}{6}}{\binom{45}{6}}$, (b) $\frac{\binom{10}{2}\binom{15}{2}\binom{20}{2}}{\binom{45}{6}}$, (c) $1-\frac{\binom{25}{6}}{\binom{45}{6}}$, (d) $\frac{\binom{10}{3}\binom{15}{3}+\binom{10}{2}\binom{15}{4}+\binom{10}{2}\binom{15}{3}\binom{20}{1}}{\binom{45}{6}}$.

32. (a) $\frac{\binom{4}{2}\binom{11}{5}}{\binom{15}{7}}$, (b) $1 - \frac{\binom{10}{7}}{\binom{15}{7}}$, (c) $\frac{\binom{9}{4}\binom{6}{3}}{\binom{15}{7}}$.

33. $1 - 6y + 15y^2 - 20y^3 + 15y^4 - 6y^5 + y^6$.

34. (a) $2\cdot 10^6$, (b) $10^6 - 9^6$, (c) $\binom{6}{3}3!$, (d) $2\mathbf{p}(9,6)$.

B.3 Chapter 3

1. 5/9, 1/6, and 5/8.

2. $1-(25/36)^{1}00-\binom{100}{99}(25/36)^{9}9(11/36)-\binom{100}{98}(25/36)^{9}8(11/36)^{2}-\binom{100}{97}(25/36)^{9}7(11/36)^{3}$.

3. 5/6

4. 9/16 and 7/9

5. (a) 0.06, (b) 0.44, (c)0.24.

6. 96/116.

7. 1/3, 1/5, and 22/30.

8. 7/400 and 3/7.

9. $\binom{52}{30}(2/5)^{3}0(3/5)^{2}2$ and $\binom{52}{50}(2/5)^{5}0(3/5)^{2}+\binom{52}{51}(2/5)^{5}1(3/5)+(2/5)^{5}2$.

10. 2/3.

12. 0.072 and 0.276.

13. 0.9.

14. (a) 0.36, (b) 0.49, (c) 0.15.

15. (a) 1/12, (b) Not independent, (c) 5/6.

16. $\binom{7}{5}(2/3)^{5}(1/3)^{2}$, $(1/3)^{7}+\binom{7}{1}(2/3)(1/3)^{6}+\binom{7}{2}(2/3)^{2}(1/3)^{5}$.

17. 1/30 and 7/10.

18. (a) Independent, (b) dependent.

19. 17/32, 1/30.

20. A and B are independent.

21. $\binom{28}{15}(1/4)^15(3/4)^13$ and $1 - [(1/4)^28 + 21(1/4)^27]$.

22. 0.4.

23. 7/16 and 3/7.

24. 1/4

25. 14/41

26. 2/3 and 5/6.

27. 0.157

28. 2/15, 28/3 = 9 and 1/3.

29. $\frac{27}{1000} + \frac{63}{900} + \frac{63}{810} + \frac{126}{720} = 0.3498$.

31. 0.3353

B.4 Chapter 4

1. −4 and 14.
2. \$0.40 and \$0.80.
3. 91/100.
4. 7 and 3.
5. $P(X = 0) = 0.1$, $P(X = 1)0.6$, $P(X = 2) = 0.3$.
6. It favors your opponent, 80 cents.
7. 2 and 7/2.
8. \$3.80.
9. 193/105 and 0.7071.

10.

-5	-4	-3	-2	-1	0	1	2	3	4	5
1/36	2/36	3/36	4/36	5/36	6/36	5/36	4/36	3/36	2/36	1/36.

11. \$2.00

12. (a) 5, (b) 9, (c) 6.

13. -45 cents.

14. $24/25 = 0.96$ and $91/100 = 0.91$

15. 393 cars.

16. (a) 73, (b) 100, (c) 35/36.

17. 7/12, \$7.00, and 3/7.

18. 15 and 68.

19. (a) 6/105, 10/105, 15/105, (b) \$6.00.

20. 210, 62.45.

21. No.

22. $210/36 = 5.83$.

23. No, your expected gain is $-1/51$.

24. 7/2

25. 7.4 cents.

26. 0 because this is a fair game.

B.5 Chapter 5

1. (a) 0.9082, (b) 0.0013, (c) 0.4972.

2. 5/8, 3/4.

3. (a) 0.1020, (b) 0.3721, (c) -1.13, (d) 0.7257, (e) 1.2, (f) 2.05.

4. 29.

5. (a) 1/3, (b) 1/4, (c) 11/18, (d) 7/12, (e) 0, (f) 1.

6. (a) 0.5328, (b) 9.4.

7. 0.0495 and 0.3983.

8. 0.9599 for $\sigma = 12$ and 0.7995 for $\sigma = 25$.

9. 1/3.

10. (a) 0.3085, (b) 0.9599, (c) 0.84, (d) 0.9876.

11. 7

12. 1/6 and 1/4.

13. (a) 0.4013, (b) 0.5468, (c) 0.8944, (d) 44, (e) 31.48, (f) 33.72.

14. (a) A, (b) A, (c) B.

15. (a) 0.3085, (b) 31, (c) 26.2.

16. 24

17. 0.772

18. 72.68

19. (a) 1/6, (b) 2/3, (c) 1/24, (d) 7/24.

20. (a) 0.2119, (b) 0.3218, (c) $P(X < 12) = 0.1151$. Yes, because about twice as many as would be expected died before they were 12 months old.

21. 38.3%

22. 0.5793 ($\sigma = 5$).

23. 115.8 seconds.

24. 98.12.

B.6 Chapter 6

1. 14, 12, and $2\sqrt{3}$.

2. 0.3911, 0.6627, 0.8775, 0.1181.

3. 0.0668.

4. (a) 0.2360, (b) 0.1738, (c) 0.1911, (d) 0.9713.

5. 10,000

6. 0.2440 and 0.4744.

7. 275/3, 3025/36, and 55/6.

8. $\mu = 210$, $\sigma = 14$, (a) 0.8944, (b) 0.4013.

9. At least 307.

10. 0.2704

11. 1,357,225

12. 1/729 and 1/27.

13. $P(X = 0) = 0.0039$, $P(X = 1) = 0.0469$, $P(X = 2) = 0.2109$, $P(X = 3) = 0.4219$, $P(X = 4) = 0.3164$.

14. (a) 0.8078 or 0.8106, (b) 0.2282.

15. 0.0573 and 0.7584.

16. 0.8944

17. (a) 0.2618, (b) $\binom{625}{127}(1/5)^{127}(4/5)^{498}+\binom{625}{128}(1/5)^{128}(4/5)^{497}+\binom{625}{129}(1/5)^{129}(4/5)^{496}$ (c) 0.1140.

18. 0.0484

19. (a) 0.2514, (b) At least 4096.

20. 0.1314 or 0.1292, and 0.0956 or 0.1034.

21. At least 827.

22. $2,280

23. 0.7051

24. 6,400 compared with 20,000 using the Law of Large Numbers.

Appendix C

Tables

TABLE I

Standard Normal Curve Areas

$\Phi(z) = P(Z \leq z)$

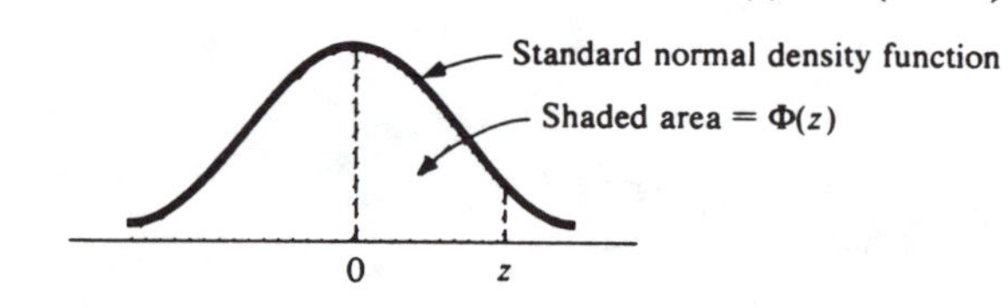

z	0.00	0.01	0.02	0.03	0.04	0.05	0.06	0.07	0.08	0.09
−3.4	0.0003	0.0003	0.0003	0.0003	0.0003	0.0003	0.0003	0.0003	0.0003	0.0002
−3.3	0.0005	0.0005	0.0005	0.0004	0.0004	0.0004	0.0004	0.0004	0.0004	0.0003
−3.2	0.0007	0.0007	0.0006	0.0006	0.0006	0.0006	0.0006	0.0005	0.0005	0.0005
−3.1	0.0010	0.0009	0.0009	0.0009	0.0008	0.0008	0.0008	0.0008	0.0007	0.0007
−3.0	0.0013	0.0013	0.0013	0.0012	0.0012	0.0011	0.0011	0.0011	0.0010	0.0010
−2.9	0.0019	0.0018	0.0017	0.0017	0.0016	0.0016	0.0015	0.0015	0.0014	0.0014
−2.8	0.0026	0.0025	0.0024	0.0023	0.0023	0.0022	0.0021	0.0021	0.0020	0.0019
−2.7	0.0035	0.0034	0.0033	0.0032	0.0031	0.0030	0.0029	0.0028	0.0027	0.0026
−2.6	0.0047	0.0045	0.0044	0.0043	0.0041	0.0040	0.0039	0.0038	0.0037	0.0036
−2.5	0.0062	0.0060	0.0059	0.0057	0.0055	0.0054	0.0052	0.0051	0.0049	0.0048
−2.4	0.0082	0.0080	0.0078	0.0075	0.0073	0.0071	0.0069	0.0068	0.0066	0.0064
−2.3	0.0107	0.0104	0.0102	0.0099	0.0096	0.0094	0.0091	0.0089	0.0087	0.0084
−2.2	0.0139	0.0136	0.0132	0.0129	0.0125	0.0122	0.0119	0.0116	0.0113	0.0110
−2.1	0.0179	0.0174	0.0170	0.0166	0.0162	0.0158	0.0154	0.0150	0.0146	0.0143
−2.0	0.0228	0.0222	0.0217	0.0212	0.0207	0.0202	0.0197	0.0192	0.0188	0.0183
−1.9	0.0287	0.0281	0.0274	0.0268	0.0262	0.0256	0.0250	0.0244	0.0239	0.0233
−1.8	0.0359	0.0352	0.0344	0.0336	0.0329	0.0322	0.0314	0.0307	0.0301	0.0294
−1.7	0.0446	0.0436	0.0427	0.0418	0.0409	0.0401	0.0392	0.0384	0.0375	0.0367
−1.6	0.0548	0.0537	0.0526	0.0516	0.0505	0.0495	0.0485	0.0475	0.0465	0.0455
−1.5	0.0668	0.0655	0.0643	0.0630	0.0618	0.0606	0.0594	0.0582	0.0571	0.0559
−1.4	0.0808	0.0793	0.0778	0.0764	0.0749	0.0735	0.0722	0.0708	0.0694	0.0681
−1.3	0.0968	0.0951	0.0934	0.0918	0.0901	0.0885	0.0869	0.0853	0.0838	0.0823
−1.2	0.1151	0.1131	0.1112	0.1093	0.1075	0.1056	0.1038	0.1020	0.1003	0.0985
−1.1	0.1357	0.1335	0.1314	0.1292	0.1271	0.1251	0.1230	0.1210	0.1190	0.1170
−1.0	0.1587	0.1562	0.1539	0.1515	0.1492	0.1469	0.1446	0.1423	0.1401	0.1379
−0.9	0.1841	0.1814	0.1788	0.1762	0.1736	0.1711	0.1685	0.1660	0.1635	0.1611
−0.8	0.2119	0.2090	0.2061	0.2033	0.2005	0.1977	0.1949	0.1922	0.1894	0.1867
−0.7	0.2420	0.2389	0.2358	0.2327	0.2296	0.2266	0.2236	0.2206	0.2177	0.2148
−0.6	0.2743	0.2709	0.2676	0.2643	0.2611	0.2578	0.2546	0.2514	0.2483	0.2451
−0.5	0.3085	0.3050	0.3015	0.2981	0.2946	0.2912	0.2877	0.2843	0.2810	0.2776
−0.4	0.3446	0.3409	0.3372	0.3336	0.3300	0.3264	0.3228	0.3192	0.3156	0.3121
−0.3	0.3821	0.3783	0.3745	0.3707	0.3669	0.3632	0.3594	0.3557	0.3520	0.3483
−0.2	0.4207	0.4168	0.4129	0.4090	0.4052	0.4013	0.3974	0.3936	0.3897	0.3859
−0.1	0.4602	0.4562	0.4522	0.4483	0.4443	0.4404	0.4364	0.4325	0.4286	0.4247
−0.0	0.5000	0.4960	0.4920	0.4880	0.4840	0.4801	0.4761	0.4721	0.4681	0.4641

z	0.00	0.01	0.02	0.03	0.04	0.05	0.06	0.07	0.08	0.09
0.0	0.5000	0.5040	0.5080	0.5120	0.5160	0.5199	0.5239	0.5279	0.5319	0.5359
0.1	0.5398	0.5438	0.5478	0.5517	0.5557	0.5596	0.5636	0.5675	0.5714	0.5753
0.2	0.5793	0.5832	0.5871	0.5910	0.5948	0.5987	0.6026	0.6064	0.6103	0.6141
0.3	0.6179	0.6217	0.6255	0.6293	0.6331	0.6368	0.6406	0.6443	0.6480	0.6517
0.4	0.6554	0.6591	0.6628	0.6664	0.6700	0.6736	0.6772	0.6808	0.6844	0.6879
0.5	0.6915	0.6950	0.6985	0.7019	0.7054	0.7088	0.7123	0.7157	0.7190	0.7224
0.6	0.7257	0.7291	0.7324	0.7357	0.7389	0.7422	0.7454	0.7486	0.7517	0.7549
0.7	0.7580	0.7611	0.7642	0.7673	0.7704	0.7734	0.7764	0.7794	0.7823	0.7852
0.8	0.7881	0.7910	0.7939	0.7967	0.7995	0.8023	0.8051	0.8078	0.8106	0.8133
0.9	0.8159	0.8186	0.8212	0.8238	0.8264	0.8289	0.8315	0.8340	0.8365	0.8389
1.0	0.8413	0.8438	0.8461	0.8485	0.8508	0.8531	0.8554	0.8577	0.8599	0.8621
1.1	0.8643	0.8665	0.8686	0.8708	0.8729	0.8749	0.8770	0.8790	0.8810	0.8830
1.2	0.8849	0.8869	0.8888	0.8907	0.8925	0.8944	0.8962	0.8980	0.8997	0.9015
1.3	0.9032	0.9049	0.9066	0.9082	0.9099	0.9115	0.9131	0.9147	0.9162	0.9177
1.4	0.9192	0.9207	0.9222	0.9236	0.9251	0.9265	0.9278	0.9292	0.9306	0.9319
1.5	0.9332	0.9345	0.9357	0.9370	0.9382	0.9394	0.9406	0.9418	0.9429	0.9441
1.6	0.9452	0.9463	0.9474	0.9484	0.9495	0.9505	0.9515	0.9525	0.9535	0.9545
1.7	0.9554	0.9564	0.9573	0.9582	0.9591	0.9599	0.9608	0.9616	0.9625	0.9633
1.8	0.9641	0.9649	0.9656	0.9664	0.9671	0.9678	0.9686	0.9693	0.9699	0.9706
1.9	0.9713	0.9719	0.9726	0.9732	0.9738	0.9744	0.9750	0.9756	0.9761	0.9767
2.0	0.9772	0.9778	0.9783	0.9788	0.9793	0.9798	0.9803	0.9808	0.9812	0.9817
2.1	0.9821	0.9826	0.9830	0.9834	0.9838	0.9842	0.9846	0.9850	0.9854	0.9857
2.2	0.9861	0.9864	0.9868	0.9871	0.9875	0.9878	0.9881	0.9884	0.9887	0.9890
2.3	0.9893	0.9896	0.9898	0.9901	0.9904	0.9906	0.9909	0.9911	0.9913	0.9916
2.4	0.9918	0.9920	0.9922	0.9925	0.9927	0.9929	0.9931	0.9932	0.9934	0.9936
2.5	0.9938	0.9940	0.9941	0.9943	0.9945	0.9946	0.9948	0.9949	0.9951	0.9952
2.6	0.9953	0.9955	0.9956	0.9957	0.9959	0.9960	0.9961	0.9962	0.9963	0.9964
2.7	0.9965	0.9966	0.9967	0.9968	0.9969	0.9970	0.9971	0.9972	0.9973	0.9974
2.8	0.9974	0.9975	0.9976	0.9977	0.9977	0.9978	0.9979	0.9979	0.9980	0.9981
2.9	0.9981	0.9982	0.9982	0.9983	0.9984	0.9984	0.9985	0.9985	0.9986	0.9986
3.0	0.9987	0.9987	0.9987	0.9988	0.9988	0.9989	0.9989	0.9989	0.9990	0.9990
3.1	0.9990	0.9991	0.9991	0.9991	0.9992	0.9992	0.9992	0.9992	0.9993	0.9993
3.2	0.9993	0.9993	0.9994	0.9994	0.9994	0.9994	0.9994	0.9995	0.9995	0.9995
3.3	0.9995	0.9995	0.9995	0.9996	0.9996	0.9996	0.9996	0.9996	0.9996	0.9997
3.4	0.9997	0.9997	0.9997	0.9997	0.9997	0.9997	0.9997	0.9997	0.9997	0.9998

TABLE II

THE BINOMIAL DISTRIBUTION

$P(X \le x)$

n	*x*	*p* = 0.05	0.10	0.15	0.20	0.25	0.30	0.35	0.40	0.45	0.50
2	0	0.9025	0.8100	0.7225	0.6400	0.5625	0.4900	0.4225	0.3600	0.3025	0.2500
	1	0.9975	0.9900	0.9775	0.9600	0.9375	0.9100	0.8775	0.8400	0.7975	0.7500
	2	1.0000	1.0000	1.0000	1.0000	1.0000	1.0000	1.0000	1.0000	1.0000	1.0000
3	0	0.8574	0.7290	0.6141	0.5120	0.4219	0.3430	0.2746	0.2160	0.1664	0.1250
	1	0.9928	0.9720	0.9392	0.8960	0.8438	0.7840	0.7182	0.6480	0.5748	0.5000
	2	0.9999	0.9990	0.9966	0.9920	0.9844	0.9730	0.9571	0.9360	0.9089	0.8750
	3	1.0000	1.0000	1.0000	1.0000	1.0000	1.0000	1.0000	1.0000	1.0000	1.0000
4	0	0.8145	0.6561	0.5220	0.4096	0.3164	0.2401	0.1785	0.1296	0.0915	0.0625
	1	0.9860	0.9477	0.8905	0.8192	0.7383	0.6517	0.5630	0.4752	0.3910	0.3125
	2	0.9995	0.9963	0.9880	0.9728	0.9492	0.9163	0.8735	0.8208	0.7585	0.6875
	3	1.0000	0.9999	0.9995	0.9984	0.9961	0.9919	0.9850	0.9744	0.9590	0.9375
	4	1.0000	1.0000	1.0000	1.0000	1.0000	1.0000	1.0000	1.0000	1.0000	1.0000
5	0	0.7738	0.5905	0.4437	0.3277	0.2373	0.1681	0.1160	0.0778	0.0503	0.0312
	1	0.9774	0.9185	0.8352	0.7373	0.6328	0.5282	0.4284	0.3370	0.2562	0.1875
	2	0.9988	0.9914	0.9734	0.9421	0.8965	0.8369	0.7648	0.6826	0.5931	0.5000
	3	1.0000	0.9995	0.9978	0.9933	0.9844	0.9692	0.9460	0.9130	0.8688	0.8125
	4	1.0000	1.0000	0.9999	0.9997	0.9990	0.9976	0.9947	0.9898	0.9815	0.9688
	5	1.0000	1.0000	1.0000	1.0000	1.0000	1.0000	1.0000	1.0000	1.0000	1.0000
6	0	0.7351	0.5314	0.3771	0.2621	0.1780	0.1176	0.0754	0.0467	0.0277	0.0156
	1	0.9672	0.8857	0.7765	0.6553	0.5339	0.4202	0.3191	0.2333	0.1636	0.1094
	2	0.9978	0.9842	0.9527	0.9011	0.8306	0.7443	0.6471	0.5443	0.4415	0.3438
	3	0.9999	0.9987	0.9941	0.9830	0.9624	0.9295	0.8826	0.8208	0.7447	0.6562
	4	1.0000	0.9999	0.9996	0.9984	0.9954	0.9891	0.9777	0.9590	0.9308	0.8906
	5	1.0000	1.0000	1.0000	0.9999	0.9998	0.9993	0.9982	0.9959	0.9917	0.9844
	6	1.0000	1.0000	1.0000	1.0000	1.0000	1.0000	1.0000	1.0000	1.0000	1.0000
7	0	0.6983	0.4783	0.3206	0.2097	0.1335	0.0824	0.0490	0.0280	0.0152	0.0078
	1	0.9556	0.8503	0.7166	0.5767	0.4449	0.3294	0.2338	0.1586	0.1024	0.0625
	2	0.9962	0.9743	0.9262	0.8520	0.7564	0.6471	0.5323	0.4199	0.3164	0.2266
	3	0.9998	0.9973	0.9879	0.9667	0.9294	0.8740	0.8002	0.7102	0.6083	0.5000
	4	1.0000	0.9998	0.9988	0.9953	0.9871	0.9712	0.9444	0.9037	0.8471	0.7734
	5	1.0000	1.0000	0.9999	0.9996	0.9987	0.9962	0.9910	0.9812	0.9643	0.9375
	6	1.0000	1.0000	1.0000	1.0000	0.9999	0.9998	0.9994	0.9984	0.9963	0.9922
	7	1.0000	1.0000	1.0000	1.0000	1.0000	1.0000	1.0000	1.0000	1.0000	1.0000
8	0	0.6634	0.4305	0.2725	0.1678	0.1001	0.0576	0.0319	0.0168	0.0084	0.0039
	1	0.9428	0.8131	0.6572	0.5033	0.3671	0.2553	0.1691	0.1064	0.0632	0.0352
	2	0.9942	0.9619	0.8948	0.7969	0.6785	0.5518	0.4278	0.3154	0.2201	0.1445
	3	0.9996	0.9950	0.9786	0.9437	0.8862	0.8059	0.7064	0.5941	0.4770	0.3633
	4	1.0000	0.9996	0.9971	0.9896	0.9727	0.9420	0.8939	0.8263	0.7396	0.6367
	5	1.0000	1.0000	0.9998	0.9988	0.9958	0.9887	0.9747	0.9502	0.9115	0.8555
	6	1.0000	1.0000	1.0000	0.9999	0.9996	0.9987	0.9964	0.9915	0.9819	0.9648
	7	1.0000	1.0000	1.0000	1.0000	1.0000	0.9999	0.9998	0.9993	0.9983	0.9961
	8	1.0000	1.0000	1.0000	1.0000	1.0000	1.0000	1.0000	1.0000	1.0000	1.0000
9	0	0.6302	0.3874	0.2316	0.1342	0.0751	0.0404	0.0207	0.0101	0.0046	0.0020
	1	0.9288	0.7748	0.5995	0.4362	0.3003	0.1960	0.1211	0.0705	0.0385	0.0195
	2	0.9916	0.9470	0.8591	0.7382	0.6007	0.4628	0.3373	0.2318	0.1495	0.0898
	3	0.9994	0.9917	0.9661	0.9144	0.8343	0.7297	0.6089	0.4826	0.3614	0.2539
	4	1.0000	0.9991	0.9944	0.9804	0.9511	0.9012	0.8283	0.7334	0.6214	0.5000
	5	1.0000	0.9999	0.9994	0.9969	0.9900	0.9747	0.9464	0.9006	0.8342	0.7461
	6	1.0000	1.0000	1.0000	0.9997	0.9987	0.9957	0.9888	0.9750	0.9502	0.9102
	7	1.0000	1.0000	1.0000	1.0000	0.9999	0.9996	0.9986	0.9962	0.9909	0.9805
	8	1.0000	1.0000	1.0000	1.0000	1.0000	1.0000	0.9999	0.9997	0.9992	0.9980
	9	1.0000	1.0000	1.0000	1.0000	1.0000	1.0000	1.0000	1.0000	1.0000	1.0000

Source: PROBABILITY AND STATISTICAL INFERENCE by Robert V. Hogg and Elliot Tanes, 1977.

n	x	p 0.05	0.10	0.15	0.20	0.25	0.30	0.35	0.40	0.45	0.50
10	0	0.5987	0.3487	0.1969	0.1074	0.0563	0.0282	0.0135	0.0060	0.0025	0.0010
	1	0.9139	0.7361	0.5443	0.3758	0.2440	0.1493	0.0860	0.0464	0.0233	0.0107
	2	0.9885	0.9298	0.8202	0.6778	0.5256	0.3828	0.2616	0.1673	0.0996	0.0547
	3	0.9990	0.9872	0.9500	0.8791	0.7759	0.6496	0.5138	0.3823	0.2660	0.1719
	4	0.9999	0.9984	0.9901	0.9672	0.9219	0.8497	0.7515	0.6331	0.5044	0.3770
	5	1.0000	0.9999	0.9986	0.9936	0.9803	0.9527	0.9051	0.8338	0.7384	0.6230
	6	1.0000	1.0000	0.9999	0.9991	0.9965	0.9894	0.9740	0.9452	0.8980	0.8281
	7	1.0000	1.0000	1.0000	0.9999	0.9996	0.9984	0.9952	0.9877	0.9726	0.9453
	8	1.0000	1.0000	1.0000	1.0000	1.0000	0.9999	0.9995	0.9983	0.9955	0.9893
	9	1.0000	1.0000	1.0000	1.0000	1.0000	1.0000	1.0000	0.9999	0.9997	0.9990
	10	1.0000	1.0000	1.0000	1.0000	1.0000	1.0000	1.0000	1.0000	1.0000	1.0000
11	0	0.5688	0.3138	0.1673	0.0859	0.0422	0.0198	0.0088	0.0036	0.0014	0.0005
	1	0.8981	0.6974	0.4922	0.3221	0.1971	0.1130	0.0606	0.0302	0.0139	0.0059
	2	0.9848	0.9104	0.7788	0.6174	0.4552	0.3127	0.2001	0.1189	0.0652	0.0327
	3	0.9984	0.9815	0.9306	0.8389	0.7133	0.5696	0.4256	0.2963	0.1911	0.1133
	4	0.9999	0.9972	0.9841	0.9496	0.8854	0.7897	0.6683	0.5328	0.3971	0.2744
	5	1.0000	0.9997	0.9973	0.9883	0.9657	0.9218	0.8513	0.7535	0.6331	0.5000
	6	1.0000	1.0000	0.9997	0.9980	0.9924	0.9784	0.9499	0.9006	0.8262	0.7256
	7	1.0000	1.0000	1.0000	0.9998	0.9988	0.9957	0.9878	0.9707	0.9390	0.8867
	8	1.0000	1.0000	1.0000	1.0000	0.9999	0.9994	0.9980	0.9941	0.9852	0.9673
	9	1.0000	1.0000	1.0000	1.0000	1.0000	1.0000	0.9998	0.9993	0.9978	0.9941
	10	1.0000	1.0000	1.0000	1.0000	1.0000	1.0000	1.0000	1.0000	0.9998	0.9995
	11	1.0000	1.0000	1.0000	1.0000	1.0000	1.0000	1.0000	1.0000	1.0000	1.0000
12	0	0.5404	0.2824	0.1422	0.0687	0.0317	0.0138	0.0057	0.0022	0.0008	0.0002
	1	0.8816	0.6590	0.4435	0.2749	0.1584	0.0850	0.0424	0.0196	0.0083	0.0032
	2	0.9804	0.8891	0.7358	0.5583	0.3907	0.2528	0.1513	0.0834	0.0421	0.0193
	3	0.9978	0.9744	0.9078	0.7946	0.6488	0.4925	0.3467	0.2253	0.1345	0.0730
	4	0.9998	0.9957	0.9761	0.9274	0.8424	0.7237	0.5833	0.4382	0.3044	0.1938
	5	1.0000	0.9995	0.9954	0.9806	0.9456	0.8822	0.7873	0.6652	0.5269	0.3872
	6	1.0000	0.9999	0.9993	0.9961	0.9857	0.9614	0.9154	0.8418	0.7393	0.6128
	7	1.0000	1.0000	0.9999	0.9994	0.9972	0.9905	0.9745	0.9427	0.8883	0.8062
	8	1.0000	1.0000	1.0000	0.9999	0.9996	0.9983	0.9944	0.9847	0.9644	0.9270
	9	1.0000	1.0000	1.0000	1.0000	1.0000	0.9998	0.9992	0.9972	0.9921	0.9807
	10	1.0000	1.0000	1.0000	1.0000	1.0000	1.0000	0.9999	0.9997	0.9989	0.9968
	11	1.0000	1.0000	1.0000	1.0000	1.0000	1.0000	1.0000	1.0000	0.9999	0.9998
	12	1.0000	1.0000	1.0000	1.0000	1.0000	1.0000	1.0000	1.0000	1.0000	1.0000
13	0	0.5133	0.2542	0.1209	0.0550	0.0238	0.0097	0.0037	0.0013	0.0004	0.0001
	1	0.8646	0.6213	0.3983	0.2336	0.1267	0.0637	0.0296	0.0126	0.0049	0.0017
	2	0.9755	0.8661	0.6920	0.5017	0.3326	0.2025	0.1132	0.0579	0.0269	0.0112
	3	0.9969	0.9658	0.8820	0.7473	0.5843	0.4206	0.2783	0.1686	0.0929	0.0461
	4	0.9997	0.9935	0.9658	0.9009	0.7940	0.6543	0.5005	0.3530	0.2279	0.1334
	5	1.0000	0.9991	0.9924	0.9700	0.9198	0.8346	0.7159	0.5744	0.4268	0.2905
	6	1.0000	0.9999	0.9987	0.9930	0.9757	0.9376	0.8705	0.7712	0.6437	0.5000
	7	1.0000	1.0000	0.9998	0.9988	0.9944	0.9818	0.9538	0.9023	0.8212	0.7095
	8	1.0000	1.0000	1.0000	0.9998	0.9990	0.9960	0.9874	0.9679	0.9302	0.8666
	9	1.0000	1.0000	1.0000	1.0000	0.9999	0.9993	0.9975	0.9922	0.9797	0.9539
	10	1.0000	1.0000	1.0000	1.0000	1.0000	0.9999	0.9997	0.9987	0.9959	0.9888
	11	1.0000	1.0000	1.0000	1.0000	1.0000	1.0000	1.0000	0.9999	0.9995	0.9983
	12	1.0000	1.0000	1.0000	1.0000	1.0000	1.0000	1.0000	1.0000	1.0000	0.9999
	13	1.0000	1.0000	1.0000	1.0000	1.0000	1.0000	1.0000	1.0000	1.0000	1.0000

Index